Student Support Materials for AQA

A-level Year 2

Physics

Sections 6, 7 and 8: Further mechanics and thermal physics, Fields and their consequences, Nuclear physics

Author: Dave Kelly

William Collins' dream of knowledge for all began with the publication of his first book in 1819.

A self-educated mill worker, he not only enriched millions of lives, but also founded a flourishing publishing house. Today, staying true to this spirit, Collins books are packed with inspiration, innovation and practical expertise. They place you at the centre of a world of possibility and give you exactly what you need to explore it.

Collins. Freedom to teach

HarperCollins Publishers
The News Building
1 London Bridge Street
London SE1 9GF

Browse the complete Collins catalogue at
www.collins.co.uk

10 9 8 7 6 5 4 3 2 1

ISBN 978-0-00-818953-2

www.collins.co.uk

A catalogue record for this book is available from the British Library

Commissioned by Gillian Lindsey
Edited by Alexander Rutherford
Project managed by Maheswari PonSaravanan at Jouve
Development by Aidan Gill
Copyedited and proof read by Janette Schubert
Typeset by Jouve India Private Limited
Original design by Hedgehog Publishing
Cover design by Angela English
Production by Lauren Crisp
Printed by CPI Group (UK) Ltd, Croydon, CR0 4YY
Cover image © Shutterstock/WHITE RABBIT83

Contents

3.6 Further mechanics and thermal physics

3.6.1 Periodic motion

3.6.1.1 Circular motion

Angular measure

You have previously studied the motion of objects which move in a straight line. You have used the equations of linear motion to describe such movement in terms of the time taken, t, the displacement, s, the acceleration, a, and the initial and final velocities, u and v.

When an object moves in a circular path, the displacement may not be important, since after each full circle the displacement is zero. It is often more useful to consider the total angle, θ, that has been turned through.

The SI unit used to measure angles is the **radian**. The radian is defined using a circle. The angle in radians, θ, at the centre of a circle is the ratio of the arc length, s, to the radius of the arc, r:

$$\theta \text{ (in radians)} = \frac{\text{arc length}}{\text{radius of arc}} \qquad \theta = \frac{s}{r}$$

Definition

One radian is the angle subtended at the centre of a circle by an arc that is equal in length to the radius.

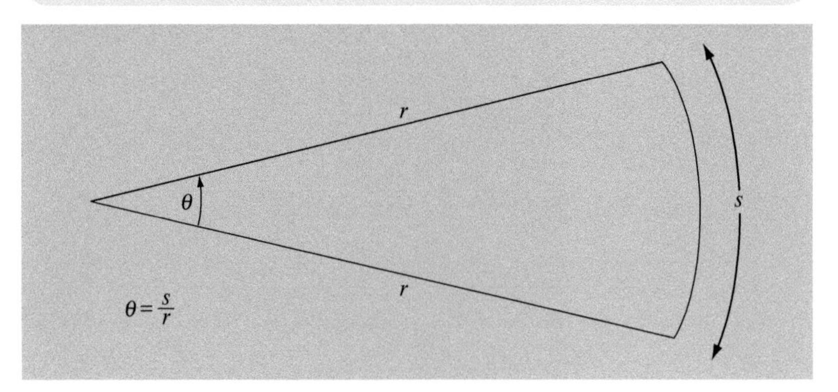

Fig 1 When the arc length is equal to the radius, the angle θ is equal to one radian.

If we consider a full circle, the arc length is the circumference, $s = 2\pi r$, so:

$$\theta \text{(radians)} = \frac{2\pi r}{r} = 2\pi$$

In other words, there are 2π radians in a full circle. As there are 360° in a full circle, we can use this to convert radians to degrees (see Table 1).

Table 1 Conversions between radians and degrees

Radians	Degrees
2π	360°
1	$360/2\pi = 57.3°$
$2\pi/360 = 0.017$	1°

Example

Calculate the angle in radians that the Earth spins through in one hour.

Answer

The Earth turns through a full circle, 2π radians, in approximately 24 hours. So in one hour:

$$\text{Angle } \theta = \frac{2\pi}{24} = 0.262\,\text{rad}$$

Rotational frequency and angular speed

The rate at which an object turns is often given in terms of the number of full circles that it completes in a given time. This is the **rotational frequency**, f, and it is typically quoted in revolutions per minute (rpm). For example, the spin speed of a washing machine may be 1000 rpm, whilst a DVD spins at between 500 and 1500 rpm. However, the SI unit of rotational frequency is the hertz, Hz, which is the number of revolutions per second.

The **angular speed**, ω, is the angle turned through in one second, measured in radians per second, rad s^{-1}. Since there are 2π radians in one revolution, the angular speed is $2\pi \times$ the angular frequency, i.e.

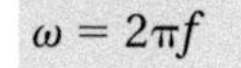

$$\omega = 2\pi f$$

Linear speed

Even though an object is moving in a circle, we can still define its 'linear' speed as the distance covered per unit time. The speed, v, of an object moving in circular motion depends on the radius of the circle, r, as well as on the angular speed, ω. For example, think of two children on a roundabout (Fig 2). They both have the same angular speed, but the child further from the centre of the roundabout (the axis of rotation) travels further in the same time, and so has a greater linear speed.

Essential Notes

All points on a rotating object have the same angular speed, but points at different distances from the axis of rotation have different linear speeds.

Fig 2
Children A and B have the same angular speed, but their linear speeds are different.

The angular speed, ω, is linked to the linear speed, v, by the equation:

$$v = r\,\omega$$

where r = radius (distance from rotational axis).
This can be derived from the definition of the radian;

$$\theta = \frac{s}{r}$$

Suppose that this is the angle turned through in a certain time, t, the angular speed, ω is:

$$\omega = \frac{\theta}{t} = \frac{s}{rt} \text{ but } \frac{s}{t} = v,$$

so $\omega = \dfrac{v}{r}$

or $$v = r\,\omega$$

Example

Find the rotational frequency and the **angular velocity** of the Earth due to its daily rotation about its axis. Find the linear speed of a point on the equator, given that the radius of the Earth is 6.4×10^6 m.

Answer

Rotational frequency is the number of revolutions per second:

$$f = \frac{1}{24 \times 60 \times 60} = 1.16 \times 10^{-5}\,\text{Hz}$$

The angular speed is given by

$$\omega = 2\pi f = 2\pi \times 1.16 \times 10^{-5} = 7.27 \times 10^{-5}\,\text{rad}\,\text{s}^{-1}$$

The speed of a point on the equator is

$$v = r\omega = 6.4 \times 10^6\,\text{m} \times 7.27 \times 10^{-5}\,\text{rad}\,\text{s}^{-1} = 465\,\text{m}\,\text{s}^{-1}$$

Centripetal acceleration

Acceleration is defined as the rate of change of velocity; this could be a change in the magnitude (the speed) *or* a change in the direction of the velocity. Since an object moving in a circular path is continuously changing the direction of its motion, it must be continuously accelerating, even if its speed is constant.

The acceleration of an object in circular motion is always directed towards the centre of the circle. It is known as **centripetal acceleration.** The magnitude of this centripetal acceleration depends on how quickly the direction is changing. In a short time, Δt, the object moves through an angle $\Delta\theta$ and the velocity changes from v_1 to v_2 (see Fig 4). The difference in velocity, Δv, can be found by subtracting v_1 from v_2. From the triangle in Fig 4, $\Delta v = v \times \Delta\theta$.

Essential Notes

Remember the difference between **vector** and **scalar** quantities. Velocity is a vector quantity; it has a direction as well as a magnitude. Speed is a scalar quantity; it is defined by a magnitude only.

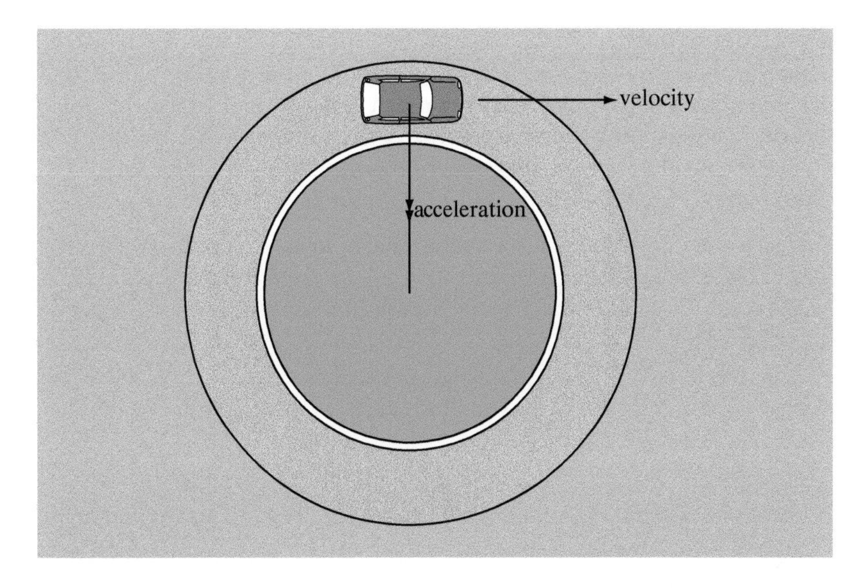

Fig 3
When a car goes round a roundabout or a corner, it is accelerating, even if it maintains a constant speed.

Notes

Objects that are travelling on a circular path may also change their speed: this results in an angular acceleration. However, work at A-level is limited to uniform circular motion, that is motion at constant speed.

The acceleration of an object moving on a circular path depends upon its speed, v, and the radius of curvature of the path. Acceleration is the rate of change of velocity. The change of velocity, Δv, in time Δt, for an object moving in a circle is $v\Delta\theta$, see Fig 4. So acceleration equals

$$a = r\omega^2 = \frac{v^2}{r}$$

Notes

The derivation of the expression for centripetal acceleration is given here for information. It will not be required in an exam.

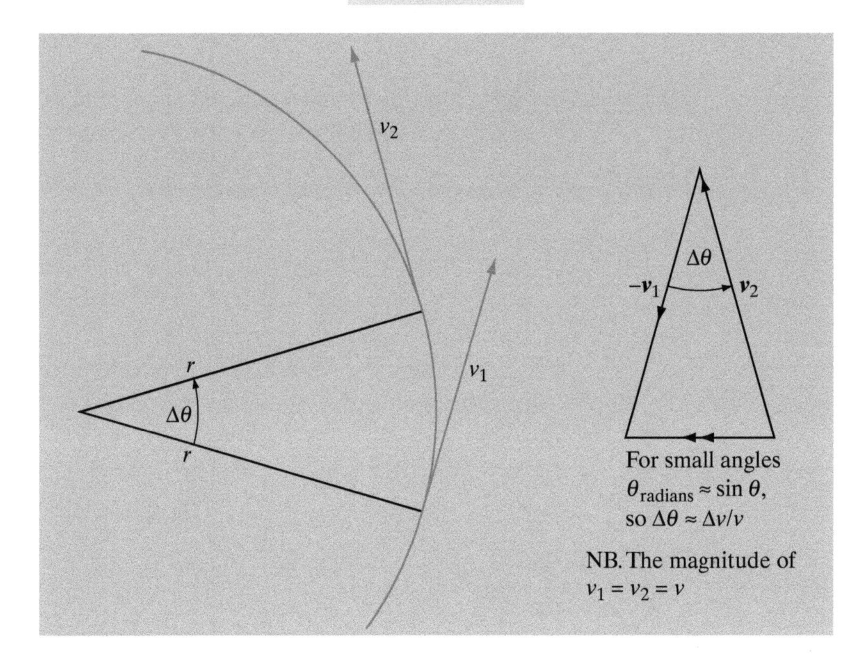

Fig 4
Deriving an expression for centripetal acceleration

Definition

The centripetal acceleration of an object moving in a circle is given by $a = r\omega^2 = v^2/r$.

Centripetal force

Since an object moving in a circle is accelerating there must be a resultant force acting on it. The resultant force could come from gravity, as in the case of a satellite orbiting the Earth, or from the tension in a string, as in the case of a conker being whirled round. The resultant force is **centripetal**; it acts towards the centre of the circle.

The size of the centripetal force that is needed to keep a mass, m, moving in a circle with a radius, r, at a velocity v, is given by Newton's Second Law in the form $F = ma$. The centripetal acceleration is v^2/r, so the centripetal force has to be

$$F = \frac{mv^2}{r} = mr\omega^2$$

Centripetal force:

- increases with mass, so that a larger force is needed to make a larger mass move at the same speed in a circle;
- increases with the square of the speed; this means that it takes four times as much friction to keep a car on the road if you take a corner twice as fast;
- decreases as the radius increases, so the force increases if the circle gets smaller.

Essential Notes

Note that the force is **centripetal** (towards the centre), NOT 'centrifugal' (away from the centre). People talk about centrifugal force 'throwing' them off their feet as a bus takes a sharp corner, or being responsible for spin-drying clothes. Using centrifugal force to explain these phenomena is a good way to make your teacher apoplectic. The person on the bus, or the water drops in a drier, are simply obeying Newton's First Law; i.e. continuing to move in a straight line, until acted on by a force.

Notes

It is a common mistake in circular motion problems to invent an *extra* force called the centripetal force. Remember that this is not an extra force but simply the resultant of the *real* forces acting on the object in circular motion. Stick to real forces, with real causes.

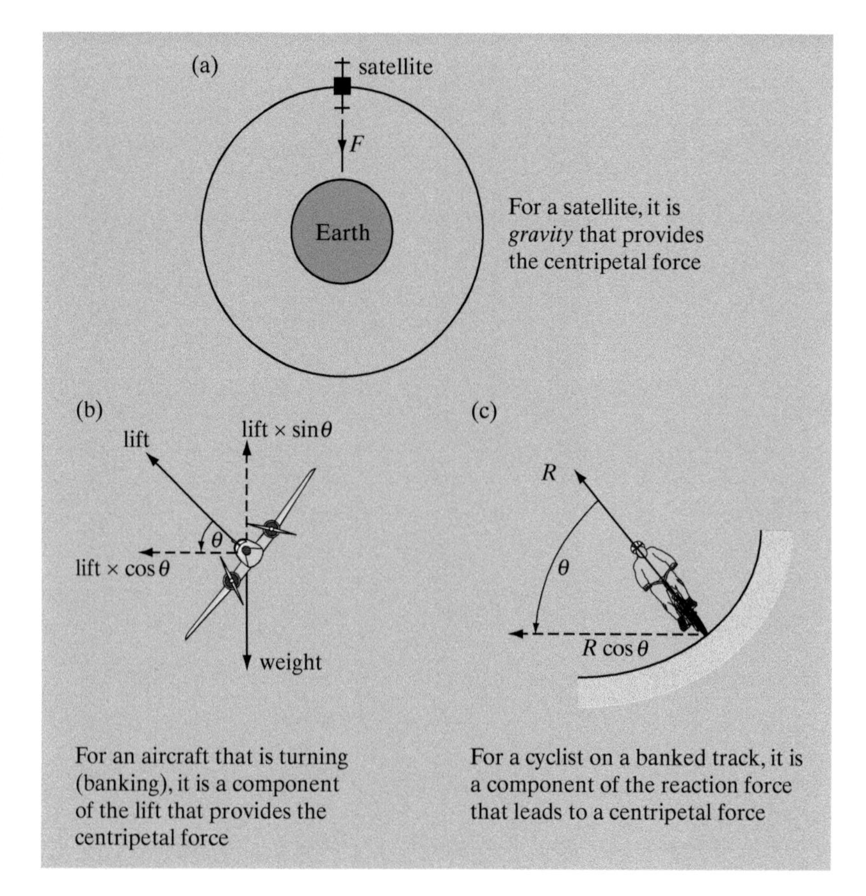

Fig 5
Examples of circular motion

When an object moves on a circular path, the force (and therefore the acceleration) are always at right angles to the velocity. In fact the force is always at right angles to the velocity. Because there is no motion in the direction of the force, there is no **work done** by the force. A satellite that is well above the Earth's atmosphere, like the Moon, can keep orbiting without any energy transfer taking place.

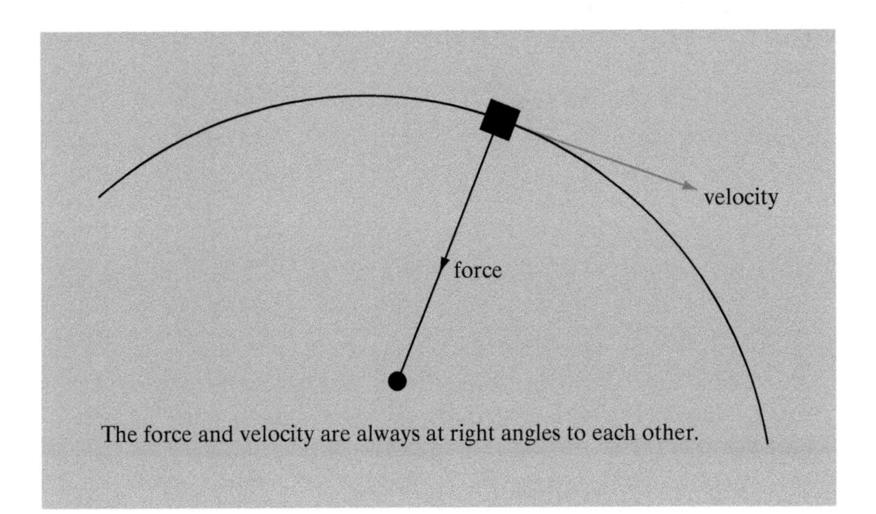

Fig 6
Force and velocity for circular motion

Example

A conical pendulum is a mass on a string that is whirled round in a horizontal circle.

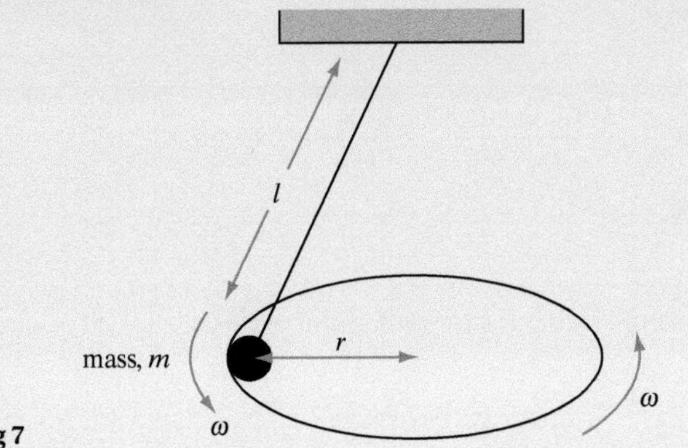

Fig 7

(a) Copy Fig 7 and mark in the forces that are acting on the mass.

(b) Identify the force or forces that are causing circular motion.

(c) If a 2 kg mass suspended on a 1 m long string is whirled around in a circle of radius 0.25 m, how fast will it be travelling?

(d) If the string broke, which direction would the mass move in at first?

Answer

(a) The forces that act on the mass are its weight, $W = mg$, and the tension, T, in the string.

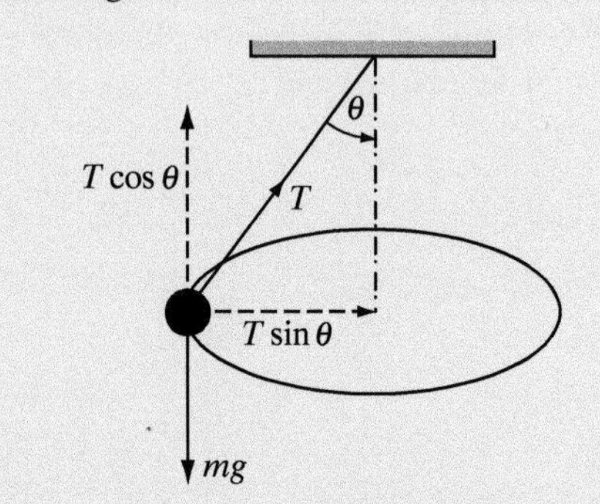

Fig 8

(b) Horizontally there is a resultant force of $T \sin \theta$ that acts towards the centre of the circle. This is the force that causes circular motion.

(c) The resultant force is $T \sin \theta = mv^2/r$.

We can find the tension by considering the vertical forces. Since the mass is moving in a horizontal circle, the vertical forces must balance: $T \cos \theta = mg$.

We can find the angle from

$$\sin \theta = \frac{r}{l} = \frac{0.25}{1} = 0.25, \quad \text{so} \quad \theta = \sin^{-1} 0.25 = 14.5°$$

Taking $g = 9.81 \text{ N kg}^{-1}$,

$$T = \frac{2\text{kg} \times 9.81\text{N kg}^{-1}}{\cos 14.5°} = 20.3\text{N}$$

This gives

$$v^2 = \frac{rT\sin\theta}{m} = \frac{0.25\text{m} \times 20.3\text{N} \times \sin 14.5°}{2\text{kg}} = 0.635\text{m}^2\text{s}^{-2}$$

The velocity is

$v = 0.797 \text{ms}^{-1}$

(d) The mass would fly off at a tangent to the circle.

Example

Jon bought nylon sewing thread, with an advertised breaking stress of 70 MPa. He wanted to test this but only had one mass: 1 kg. He tied the mass to the thread and whirled it around in a horizontal circle, gradually increasing the speed of rotation until the thread broke.

(a) Explain how Jon's method could lead to an estimate of breaking stress.

(b) Estimate how fast the mass was moving when the thread broke.

(c) Apart from timing the rotations, can you suggest a way for Jon to get an estimate of the velocity of the mass, when the thread breaks?

Answer

(a) The tension in the thread provides the centripetal force = $mr\omega^2$. The tension will increase (proportional to ω^2) as the mass orbits faster. The force required to break the thread is: = $70 \times 10^6 \times$ cross-sectional area. We need to estimate the radius of the thread, say 0.5 mm.

The breaking force = $\pi r^2 \times 70 \times 10^6 = 55$ N. This would occur at $\omega = \sqrt{(F / mR)}$, where R is the radius of the circle, say 1 m.

(b) This gives $\omega = \sqrt{(55 / 1 \times 1)} = 7.4$ rads^{-1} = 1.2 rev per second or 70 rpm. The mass would be travelling at a linear speed of $v = r\,\omega = 1 \times 7.4$ m s^{-1} = 7.4 m s^{-1}.

(c) Jon could estimate the speed of the mass at the instant the thread broke, by treating it as a projectile and measuring the range. $s = ut$ (horizontally) and find t from the vertical distance fallen, $s = \frac{1}{2} at^2$.

3.6.1.2 Simple harmonic motion (SHM)

A car bouncing on its suspension, a child on a playground swing and the vibrations of a water molecule are all examples of **oscillations**. An oscillation is a repetitive, to-and-fro motion about a fixed position. This sort of motion is caused by a resultant force that is always directed to the same point. This resultant force, which changes direction as the object oscillates, is referred to as a **restoring force**.

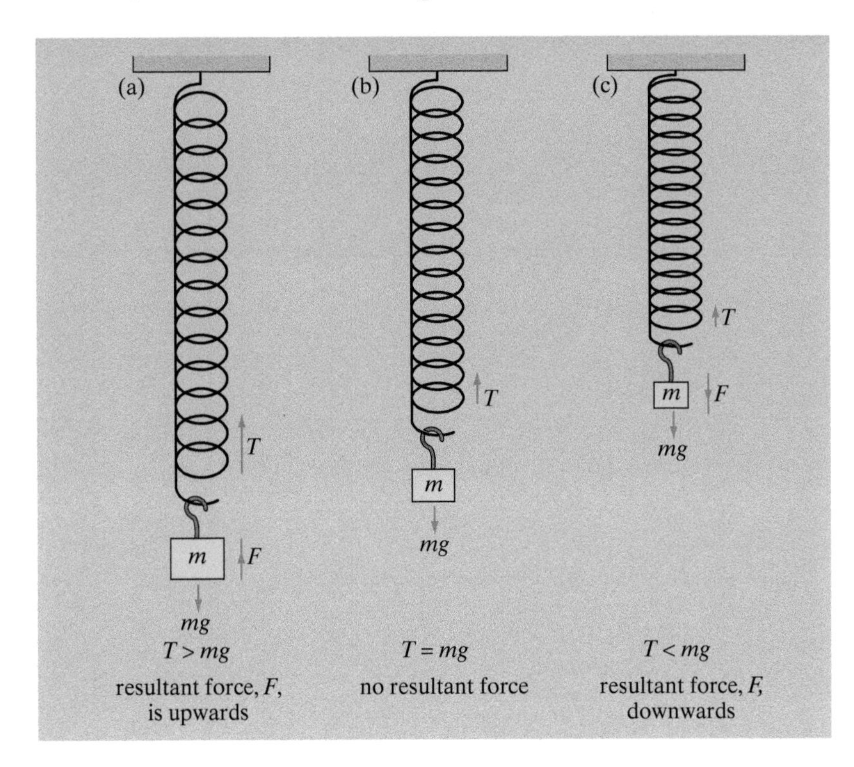

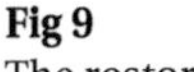

Fig 9
The restoring force on a mass suspended from a spring

Simple harmonic motion, or **SHM**, is a special case of oscillatory motion. If the magnitude of the restoring force on a body is proportional to its distance from the equilibrium position, we say that it moves with simple harmonic motion. Assuming that the mass of the object doesn't change, the acceleration will follow the same pattern.

Essential Notes

The negative sign arises because the acceleration and the displacement are always in opposite directions, so that a positive displacement causes a negative acceleration, and vice versa.

Definition

The acceleration, a, of an object moving with simple harmonic motion, is always proportional to its displacement, x, from a fixed point. The acceleration is always directed towards that fixed point.

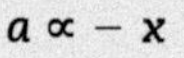

$$a \propto -x$$

Fig 10
Acceleration against displacement for an object moving with SHM

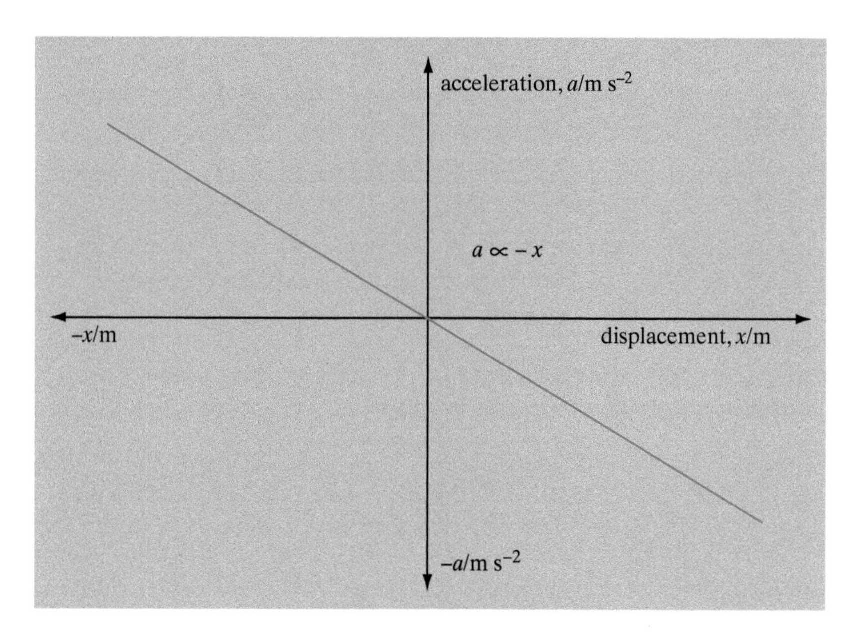

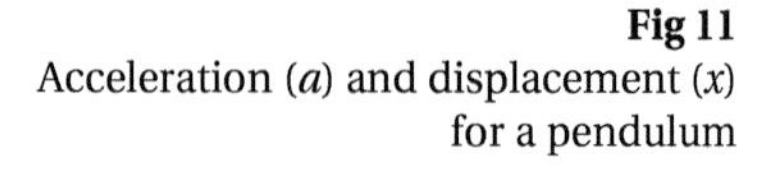

Fig 11
Acceleration (a) and displacement (x) for a pendulum

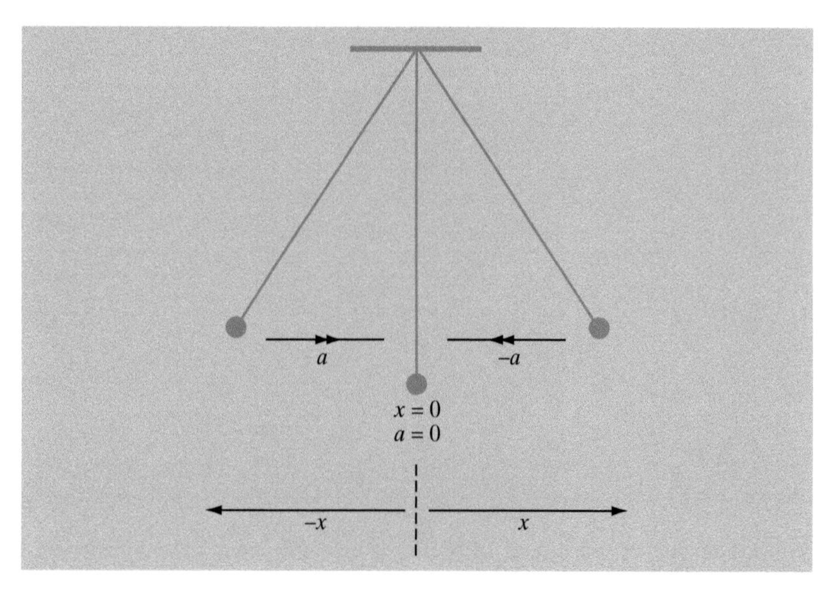

Essential Notes

It can be difficult to realise that the pendulum has its greatest acceleration when it isn't moving at all. Conversely, the pendulum has no acceleration when it is moving fastest.

A swinging pendulum is an example of a body moving with SHM. The restoring force, and the acceleration, is greatest when the pendulum is furthest from equilibrium, when the **displacement** is equal to its maximum value or **amplitude**, A. The acceleration drops to zero when the displacement of the pendulum is zero.

Definitions

***Displacement** is the distance from a fixed point in a certain direction. It is a vector quantity.*

***Amplitude** is the magnitude of the maximum displacement. It is a scalar quantity.*

Period, frequency and acceleration

Oscillations are often referred to as **periodic motion**. The vibrations of a guitar string or the motion of a piston in a car engine are examples of periodic motion. In each case a pattern of motion is repeated over and over again. The time taken to complete one full cycle of motion is called the **period**, T.

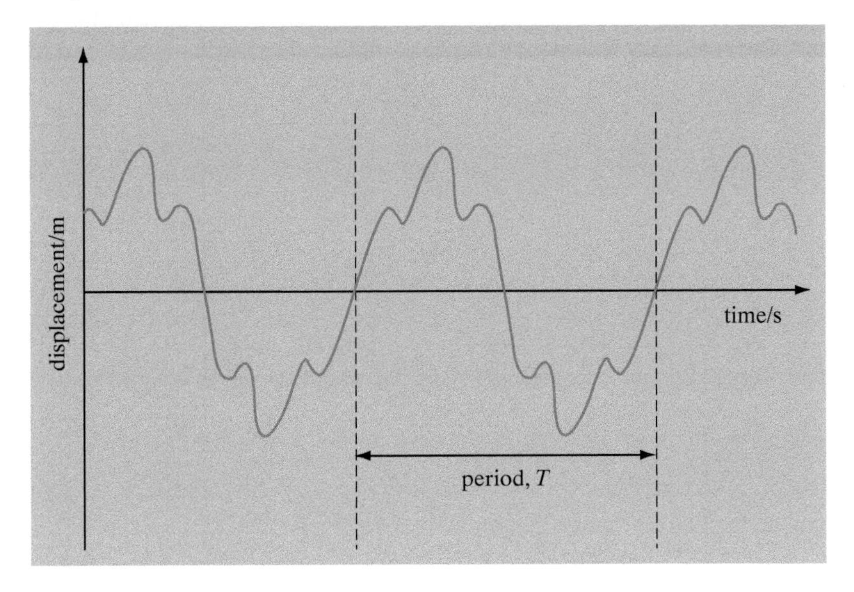

Fig 12
Periodic motion

The number of oscillations that an object completes in one second is called the **frequency** of the oscillation. Frequency is measured in hertz, Hz. Frequency is related to period by the expression:

$$\text{frequency} = \frac{1}{\text{period}} \qquad f = \frac{1}{T}$$

High-frequency oscillations involve high accelerations, since the body has to speed up and slow down many times per second. For simple harmonic oscillations the acceleration is linked to the frequency and the displacement by the equation

$$a = -(2\pi f)^2 x$$

Notes

Remember from circular motion that the angular speed is given by $\omega = 2\pi f$.

Example

The piston in a car engine moves with a motion that is approximately simple harmonic. One cycle of motion takes 0.017 s and the piston moves through a total distance of 100 mm. Calculate the maximum acceleration of the piston.

Answer

The piston's frequency is: $f = \dfrac{1}{0.017\,\text{s}} = 58.8\,\text{Hz}$

The period of the motion is 0.017 s. The piston moves through a maximum displacement of 100 mm. This will be equal to 50 mm (half the total movement of the piston), so

$$a = -\omega^2 x - (2\pi f)^2 x = -(2\pi \times 58.8)^2 \times 0.05 = 6800\,\text{ms}^{-2} \text{ (to 2 s.f.)}$$

in a direction opposite to that of the displacement.

Displacement and time

Bodies that move with simple harmonic motion have a displacement that varies sinusoidally with time.

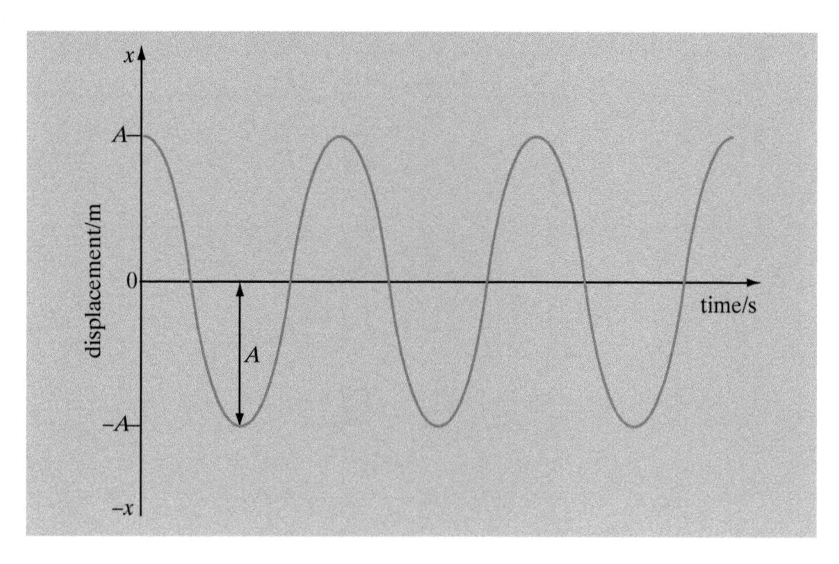

Fig 13
Displacement against time for a simple harmonic oscillator

Essential Notes

To convert degrees to radians you need to remember that there are 2π radians in a circle. So 2π radians = 360°. From this you can work out that 1 radian = $360/2\pi$ = 57.3°. See Table 1, page 1.

The equation that describes the graph in Fig 13 is

$$x = A\cos(\omega t) = A\cos(2\pi ft)$$

where x is the displacement, A is the amplitude, f is the frequency and t is the time. The term $2\pi ft$ has units of radians (see page 1).

Example

(a) Plot a displacement–time graph to show three cycles of an oscillation that has an amplitude of 6 mm and a frequency of 5000 Hz.

(b) Calculate the displacement after 0.24 ms.

Answer

(a) You need to use the equation

$$x = A\cos(\omega t) = A\cos(2\pi f t)$$

$$x = 0.006\cos(10000\pi t)$$

Now you need to choose suitable values of t to calculate values of x for plotting. Values of $t = 1, 2, 3$, etc. are too large and represent points which are thousands of oscillations apart.

The frequency of the oscillation is 5000 Hz, so the oscillation has a period of $1/f = 1/5000 = 0.0002$ s or 0.2 ms.

The oscillation will have a maximum displacement at 0, 0.2, 0.4, 0.6 ms etc.

The oscillation will have a minimum (negative) displacement at 0.1, 0.3, 0.5 ms etc.

The oscillation will have zero displacement in between these points, at 0.05, 0.15, 0.25, 0.35 ms, etc.

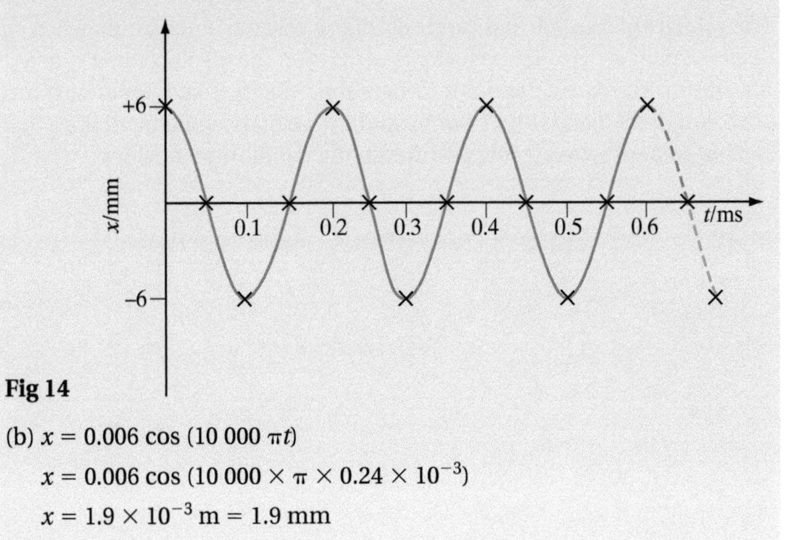

Fig 14

(b) $x = 0.006 \cos(10\,000\,\pi t)$

$x = 0.006 \cos(10\,000 \times \pi \times 0.24 \times 10^{-3})$

$x = 1.9 \times 10^{-3}$ m $= 1.9$ mm

Notes

Remember to set your calculator to radians when you calculate this formula.

Velocity and time

The velocity of a particle that is moving with SHM can be found from the gradient of the displacement vs time graph.

Essential Notes

Remember (from AS/A-Level Year 1, Section 4) that velocity $= \Delta x/\Delta t$.

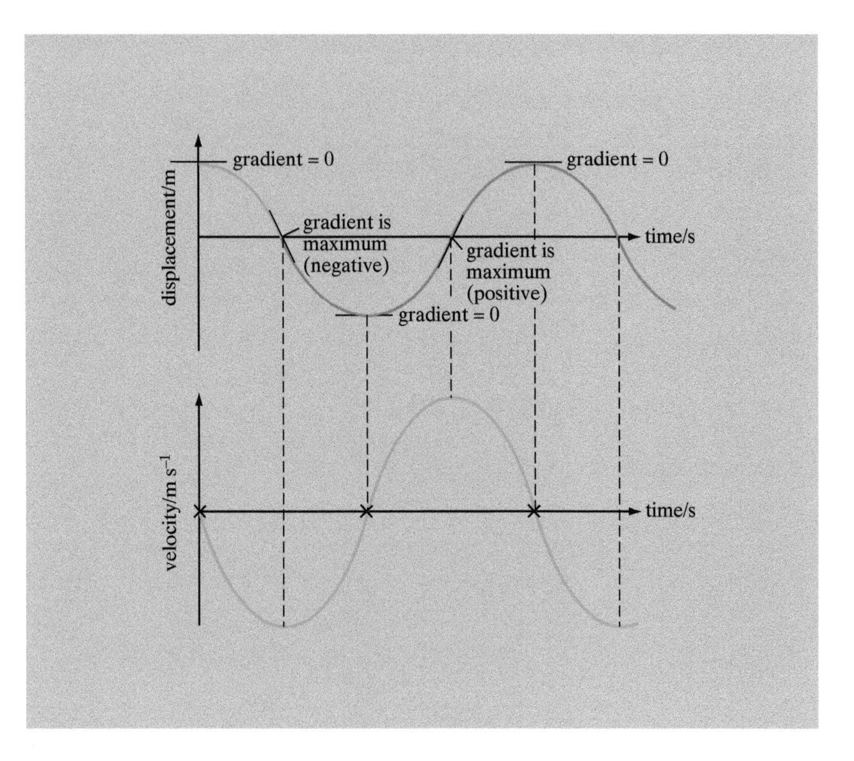

Fig 15
Displacement and velocity vs time for a simple harmonic oscillator

The velocity of a simple harmonic oscillator reaches a maximum when the displacement is zero. The velocity is zero when the displacement is at a maximum. Think of a pendulum where the velocity is zero for an instant at each end of the pendulum's swing, and the velocity of the pendulum bob is at its greatest when it swings through the equilibrium position.

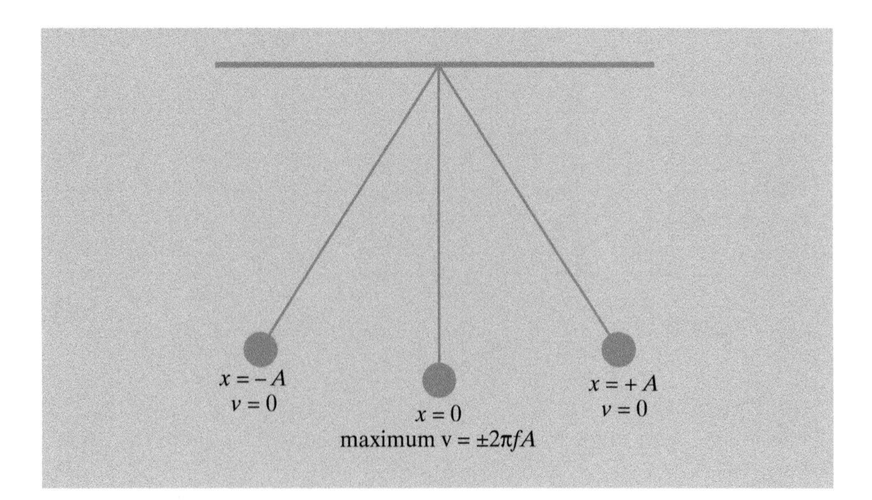

Fig 16
Velocity and displacement for a pendulum bob

This pattern of changing velocity can be linked to the displacement by the equation

$$v = \pm 2\pi f\sqrt{A^2 - x^2}$$

so that the maximum velocity, v_{max} occurs at $x = 0$:

$$v_{max} = \pm\, \omega A = \pm 2\pi fA$$

Example

A mass suspended from a spring oscillates with a period of 1.5 s. The amplitude of the oscillations is 5 cm. Find

(a) the maximum velocity

(b) the minimum velocity

(c) the velocity when the displacement is half the amplitude

(d) the displacement when the velocity is half of its maximum value.

Answer

The period of the oscillation is 1.5 s, so the frequency is

$$f = \frac{1}{1.5} = 0.67\,\text{Hz}$$

(a) The maximum velocity occurs when there is no displacement, $x = 0$:

$$v = \pm 2\pi f\sqrt{A^2 - x^2} = \pm 2\pi f A$$
$$= \pm 2\pi \times 0.67 \times 0.05 = 0.21\,\text{m}\,\text{s}^{-1}$$

(b) The minimum velocity is when $x = A$, and then $v = 0\ \text{m s}^{-1}$.

(c) The velocity at $x = 0.025$ m is

$$v = \pm 2\pi \times 0.67 \times \sqrt{0.05^2 - 0.025^2} = \pm 0.18\,\text{m}\,\text{s}^{-1}$$

(d) When the velocity is $\frac{1}{2}$ of its maximum value, $v = 0.105\ \text{m s}^{-1}$.

$$v^2 = 4\pi^2 f^2(A^2 - x^2) \quad \text{so} \quad x^2 = A^2 - \frac{v^2}{4\pi^2 f^2}$$

This gives $x = \pm 0.043$ m from equilibrium.

Essential Notes

Velocity is not proportional to displacement. So, for example, the velocity at $x = \frac{1}{2}$ A does not equal $\frac{1}{2}\, v_{max}$.

Essential Notes

Remember that the instantaneous velocity is the gradient of a displacement–time graph, $\Delta x/\Delta t$, and that the instantaneous acceleration is the gradient of a velocity-time graph, $\Delta v/\Delta t$.

Acceleration

Acceleration is the rate of change of velocity, so an oscillating object will have its greatest acceleration when the velocity is changing most quickly, at the ends of the oscillation.

$$\text{maximum acceleration} = \omega^2 A$$

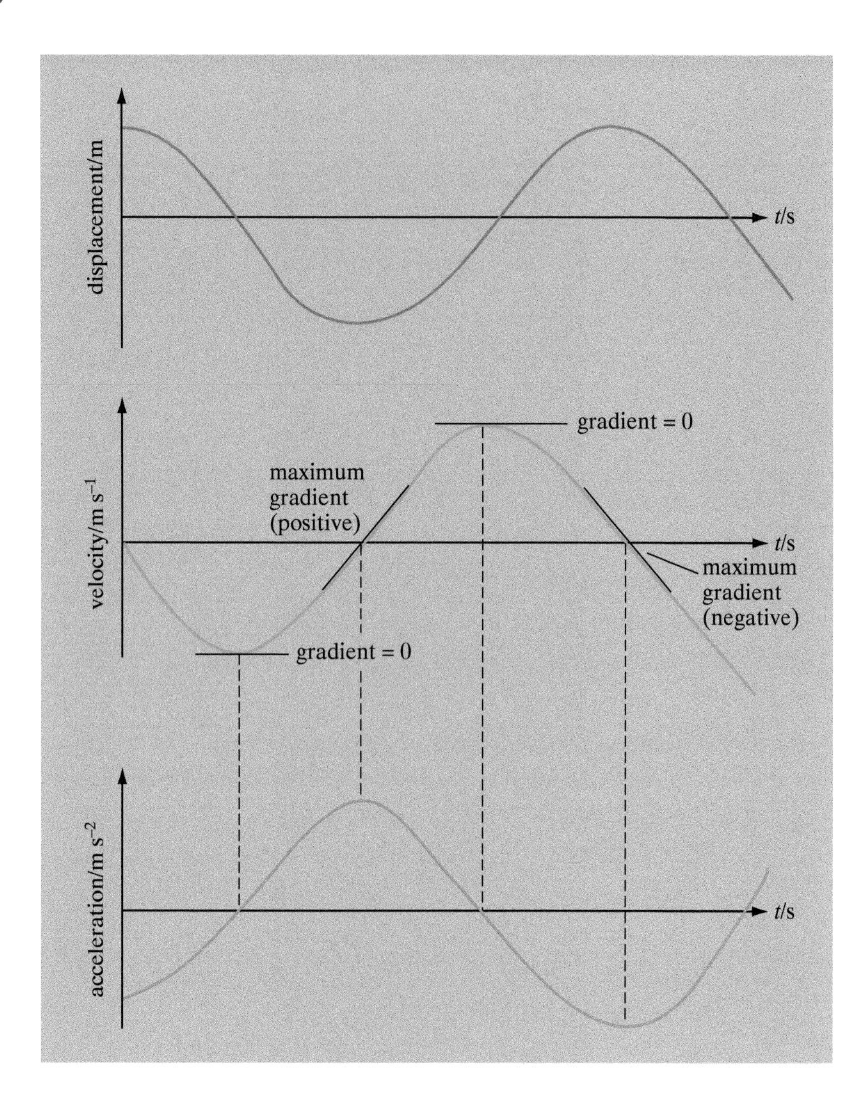

Fig 17
Displacement, velocity and acceleration vs time graph for a simple harmonic oscillator

The magnitude of the acceleration reaches its maximum value at the same time as the displacement, though they are in opposite directions.

Example

A loudspeaker is vibrating at 250 Hz with an amplitude of 1 mm. If the vibrations change to a frequency of 500 Hz and an amplitude of 2 mm;

(a) What would happen to the maximum velocity?

(b) What would happen to the maximum acceleration?

Answer

(a) $v_{max} = 2\pi fA$. Since A and f are both doubled $v_{max} \rightarrow 4\times$ its original value.

(b) $a_{max} = (2\pi f)^2 A$. Since A is doubled and f^2 goes to $4\times$ its original value, the maximum acceleration increases by a factor of $\times 8$.

3.6.1.3 Simple harmonic systems

Mass on a spring

A mass bouncing on a spring is another example of simple harmonic motion.

To show that this is SHM we need to consider the forces acting on the mass.

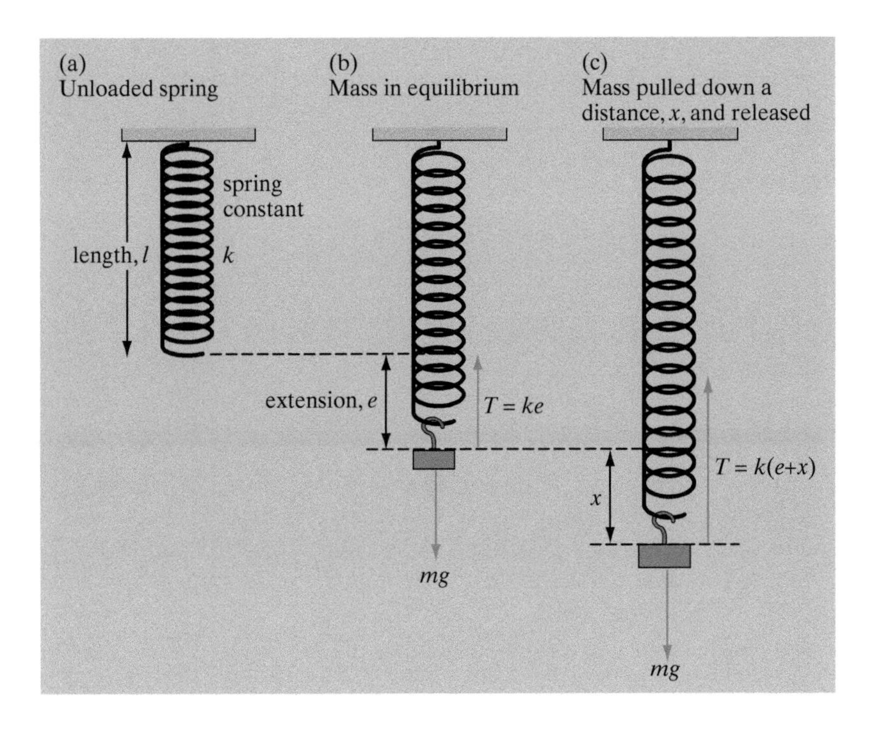

Fig 18
The forces acting on a mass on a spring

When the mass is in equilibrium, the sum of the forces must be zero. So, taking downward as the positive direction:

$$W - T = 0 \quad \text{or} \quad mg - ke = 0$$

When the mass is pulled down by a further displacement, x, the sum of the forces is equal to the mass × acceleration (from Newton's Second Law):

$$mg - k\,(e + x) = ma,$$

but $mg = ke$, so:

$$-kx = ma \quad \text{or} \quad a = -\left(\frac{k}{m}\right)x$$

For any given mass and spring, m and k are constants, so the acceleration is proportional to the displacement and directed in the opposite direction. The motion is therefore simple harmonic.

Essential Notes

The negative sign shows that the net force is upwards, whereas the displacement x is measured from the equilibrium position in the downwards direction.

If we compare the equation for acceleration with the general SHM equation, we can find an expression for the period of the oscillations.

Mass–spring system $a = -\left(\frac{k}{m}\right)x$

General SHM equation $a = -(2\pi f)^2 x$

This gives $\frac{k}{m} = (2\pi f)^2$

So $f = \frac{1}{2\pi}\sqrt{\frac{k}{m}}$

As $T = 1/f$,

$$T = 2\pi\sqrt{\frac{m}{k}}$$

The period of a mass–spring system depends on the mass and the spring constant.

Example

A family car has a mass of 1000 kg when it is not loaded. This mass is supported equally by four springs. When the car is fully loaded its mass goes up to 1250 kg and the springs compress by a further 2 cm. When the car goes over a bump in the road, it bounces on its springs. Find the period of these oscillations.

Answer

The formula for the period of a mass–spring system is $T = 2\pi\sqrt{\frac{m}{k}}$.

We know the mass of the system but we need to calculate the spring constant, k.

The extra weight of 250 kg × 9.81 N kg^{-1} = 2450 N, will depress the four springs by 0.02 m.

If each spring carries $\frac{1}{4}$ of the weight, we can use Hooke's Law, $F = ke$, to find the spring constant of one spring:

$$k = \frac{F}{e} = \frac{613}{0.02} = 3.06 \times 10^4\,\text{N m}^{-1}$$

Then we can find the period of oscillation of a spring. The effective mass oscillating on each spring is 1250/4 = 313 kg.

$$T = 2\pi\sqrt{\frac{m}{k}} = 2\pi\sqrt{\frac{313}{3.06 \times 10^4}} = 0.64\text{s}$$

Fig 19
Simple pendulum

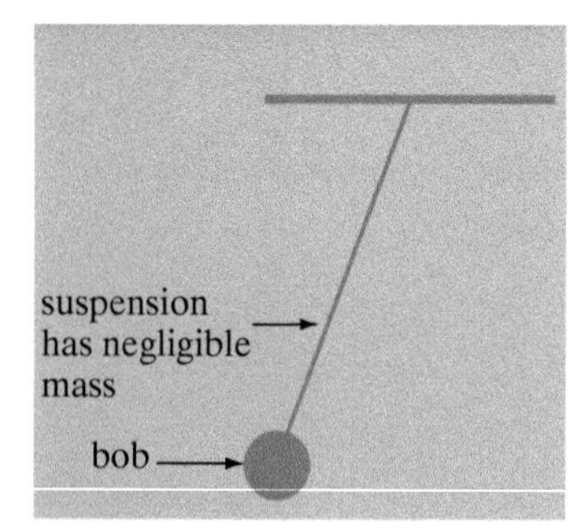

The simple pendulum

A 'simple' pendulum has all its mass concentrated at the free end. This mass is called the pendulum 'bob'. The mass of a simple pendulum's suspension is small compared to the bob's mass and so we can neglect it. When this is not true, the pendulum is referred to as a compound pendulum.

We can show that an oscillating simple pendulum is an example of simple harmonic motion. To do this we need to prove that the acceleration is proportional to the displacement from equilibrium, and is directed in the opposite direction to the displacement. First we need to consider the forces acting on the pendulum. There are only two forces acting, the weight of the bob, W, and the tension in the string, T. We resolve the weight into two components, acting parallel and perpendicular to the tension (Fig 20).

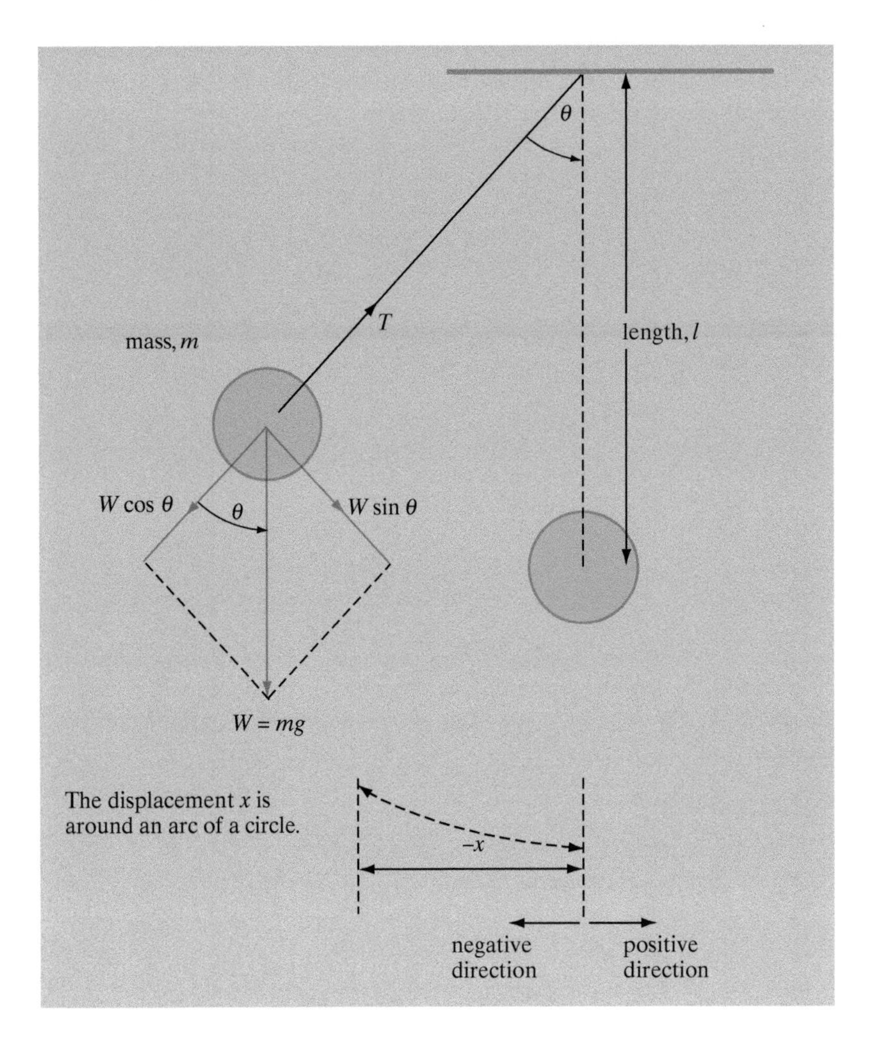

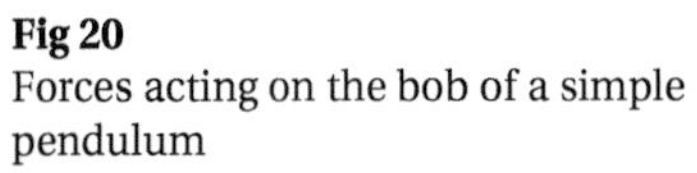

Fig 20
Forces acting on the bob of a simple pendulum

Essential Notes

If the angle θ is small enough, we can assume that θ in radians is approximately equal to $\sin\theta$. (This is valid up to about 10°, $\sin 10° = 0.17364$ and θ (radians) $= 0.17453$, a difference of only about 0.5%.)

$W\sin\theta$ is the only force acting perpendicular to the string. It causes an acceleration towards the equilibrium position.

$$W\sin\theta = ma \qquad \text{or} \qquad mg\sin\theta = ma$$

$$a = g\sin\theta$$

Essential Notes

The negative sign appears because we have defined displacement to the left as negative, and a negative displacement produces an acceleration to the right.

If the angle θ is small, we can say that $\sin\theta \cong \theta = -x/l$.

So

$$a = -\left(\frac{g}{l}\right)x$$

The definition of SHM states that the acceleration must be proportional to the displacement, and is directed in the opposite direction. Since g/l is constant for this pendulum, we can say that acceleration is proportional to displacement and the motion is therefore simple harmonic. This is only true for small angles of swing.

We can find an expression for the period of the motion by comparing the acceleration with the general SHM equation.

Simple pendulum $\quad a = -\left(\frac{g}{l}\right)x$

General SHM equation $\quad a = -(2\pi f)^2 x$

We can see that $\quad \frac{g}{l} = (2\pi f)^2$

So $\quad f = \frac{1}{2\pi}\sqrt{\frac{g}{l}}$

As $T = 1/f$,

$$T = 2\pi\sqrt{\frac{l}{g}}$$

The period of a simple pendulum just depends on its length and on the acceleration due to gravity. The mass of a pendulum does not affect its period.

Example

A child on a playground swing oscillates with a motion that is approximately simple harmonic.

(a) If the child has a mass of 30 kg and the swing is 2 m long, calculate the period of oscillation.

(b) The child's younger sister, mass 20 kg, now gets onto the swing instead. What difference would you expect to the period of the pendulum?

Answer

(a) We use the equation $T = 2\pi\sqrt{\frac{l}{g}}$

$$T = 2\pi\sqrt{\frac{2}{9.81}} = 2.8\,\text{s (to 2 s.f.)}$$

(b) The period will be almost unchanged. The period of a pendulum does not depend on its mass. The effective length of the pendulum would be altered if its centre of mass was different, and this would affect the period slightly.

Energy in oscillations

Objects that oscillate are continually transferring energy from potential energy to kinetic energy and back again. A pendulum uses its kinetic energy to do work against gravity, storing gravitational potential energy. This energy is then transferred as kinetic energy as the pendulum accelerates back towards the centre of the oscillation.

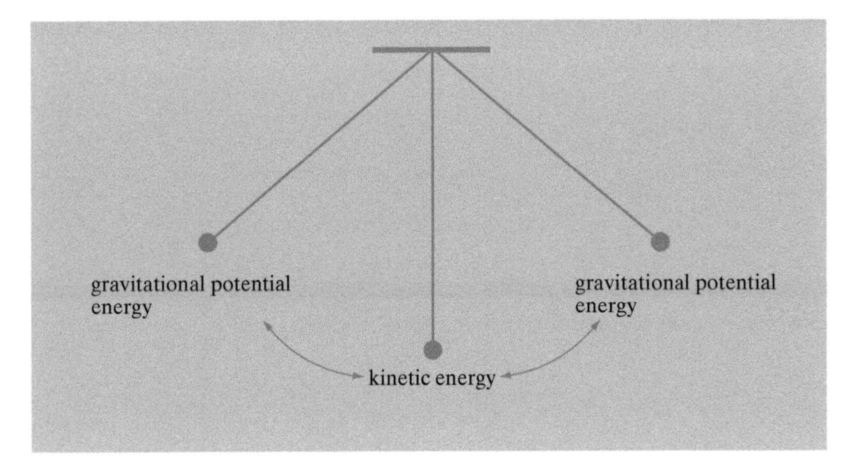

Fig 21
Energy transfer during the oscillation of a pendulum

If the oscillating system does not transfer energy to the surroundings, the total energy remains constant:

Total energy = kinetic energy + potential energy

The variation of kinetic and potential energy with displacement is shown in Fig 22.

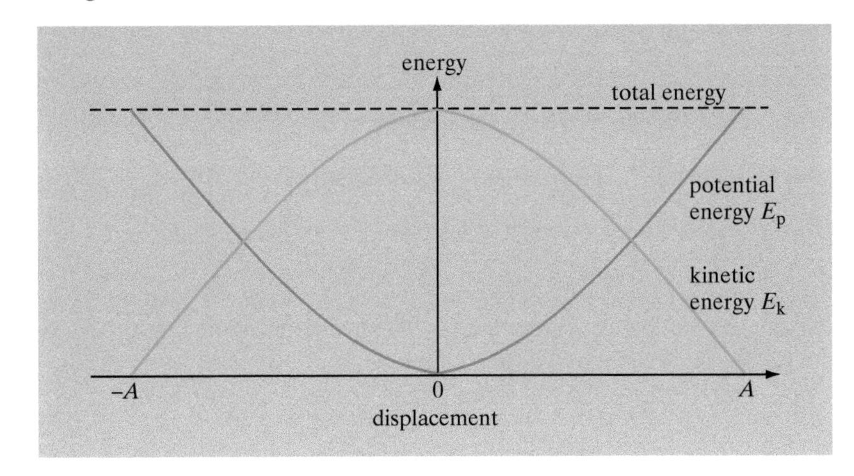

Fig 22
Energy vs displacement for a pendulum oscillation

Fig 23
A horizontal mass–spring system

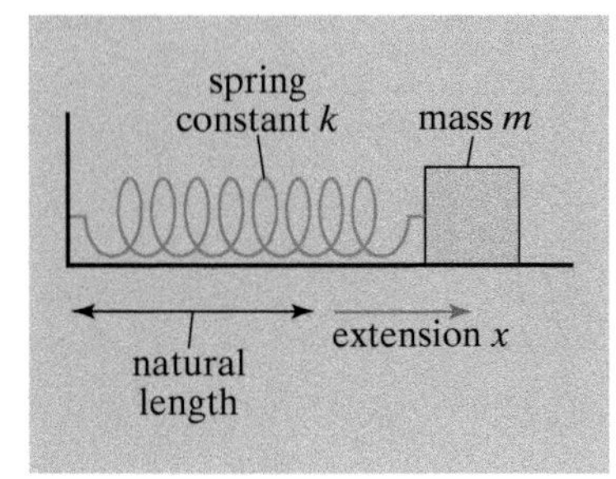

Horizontal mass–spring system

We can investigate energy changes in a simple harmonic oscillator by considering a mass on a horizontal, frictionless surface. The mass is connected to a spring which extends and contracts. We assume that the spring obeys Hooke's law in compression as well as extension (Fig 23).

The mass is pulled to the right, to a maximum value of displacement, A, and then released. The mass oscillates in simple harmonic motion about the equilibrium position, compressing and extending the spring. At each end of the oscillation, when $x = \pm A$, the system has potential energy, stored as elastic strain energy in the spring. As the mass passes through the equilibrium position, when the spring is at its natural length, all the energy is kinetic.

The kinetic energy of the mass is given by

$$E_k = \tfrac{1}{2} mv^2$$

Since this is SHM, $v^2 = 4\pi^2 f^2(A^2 - x^2)$, so

$$\begin{aligned} E_k &= 2m\pi^2 f^2(A^2 - x^2) \\ &= 2m\pi^2 f^2 (A^2 - A^2\cos^2 2\pi ft) \\ &= 2m\pi^2 f^2 A^2 (1 - \cos^2 2\pi ft) \end{aligned}$$

Notes

Remember from AS/A-Level Year 1, Section 4 that the energy stored in a spring is $\frac{1}{2}$ $F\Delta L$, where ΔL is the extension, and since Hooke's Law gives $F = k\Delta L$, the energy stored is equal to $\frac{1}{2}k\Delta L^2$.

Since $1 - \cos^2\theta = \sin^2\theta$, the kinetic energy is given by:

$$E_k = 2m\pi^2 f^2 A^2 \sin^2 2\pi ft$$

The potential energy is stored in the spring as elastic strain energy:

$$E_p = \tfrac{1}{2} kx^2$$

where k is the spring constant. For SHM, $x = A \cos 2\pi ft$, so

$$E_p = \tfrac{1}{2}kA^2 \cos^2 2\pi ft$$

The potential energy is therefore proportional to $\cos^2 2\pi ft$.

The total energy, E_{Total}, of the mass–spring system is:

$$\begin{aligned} E_{Total} &= E_k + E_p \\ &= 2m\pi^2 f^2 A^2 \sin^2(2\pi ft) + \tfrac{1}{2} kA^2 \cos^2(2\pi ft) \end{aligned}$$

Since the period, T, of a mass–spring system is given by:

$T = 2\pi\sqrt{(m/k)}$, the frequency, f, is given by:

$f = (\frac{1}{2}\pi)\sqrt{(k/m)}$. So $4\pi^2 f^2 m = k$.

So we can write the equation for total energy in terms of the mass, the amplitude and the frequency of oscillations:

$$\begin{aligned} E_{Total} = E_k + E_p &= 2m\pi^2 f^2 A^2 \sin^2(2\pi ft) + \tfrac{1}{2} 4\pi^2 f^2 mA^2 \cos^2(2\pi ft) \\ &= 2m\pi^2 f^2 A^2\{\sin^2(2\pi ft) + \cos^2(2\pi ft)\} \\ &= 2m\pi^2 f^2 A^2 \end{aligned}$$

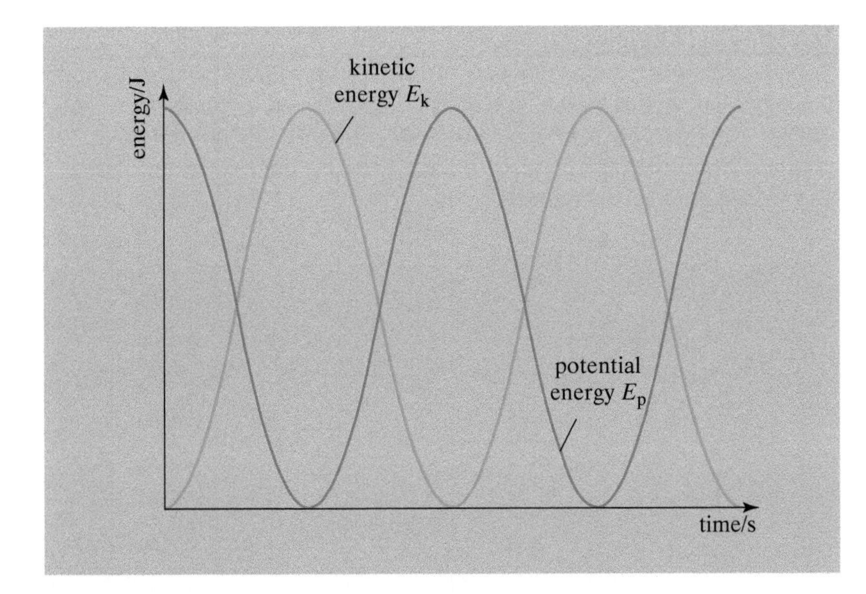

Fig 24
The variation of kinetic and potential energy with time in simple harmonic oscillations

Example

A mass of 500 g is hung from a spring which extends by 6 cm. The mass is pulled down a further 4 cm and then released. Calculate the maximum kinetic energy of the mass, and hence find its maximum velocity.

Answer

The energy stored in a stretched spring is $\frac{1}{2}kx^2$, where k is the spring constant.

$$k = \frac{\text{force}}{\text{extension}} = \frac{4.91\text{N}}{0.05} = 81.8\text{N m}^{-1} \text{ (taking } g \text{ as } 9.81\text{N kg}^{-1}\text{)}$$

The elastic potential energy in the spring when it is extended by a further 4 cm (total extension = 10 cm) is:

$$E = \tfrac{1}{2} \times 81.8 \times 0.10^2 = 0.409\text{J}$$

This is transferred as kinetic energy and gravitational potential energy when the spring is released. The maximum kinetic energy of the mass will occur when it passes through the equilibrium position. When the mass passes back through the equilibrium position, the gravitational potential energy gained is

$$\Delta E_p = mg\,\Delta h = 0.5 \times 9.81 \times 0.04 = 0.196\text{J}$$

The energy still stored in the spring is $E = \frac{1}{2} \times 81.8 \times 0.06^2 = 0.147\text{J}$

Therefore the kinetic energy is the remainder:

$$E_k = 0.409\text{ J} - 0.196\text{ J} - 0.147\text{ J} = 0.066\text{ J}$$

Because kinetic energy is $E_k = \frac{1}{2}mv^2$, the velocity is

$$v = \sqrt{\frac{2E_k}{m}} = 0.51\text{m s}^{-1}$$

Damped oscillations

In real oscillating systems there are always resistive forces, such as friction or air resistance, that lead to energy being transferred to the surroundings. These oscillations are said to be **damped**. The amplitude of a damped oscillation gradually decreases as the total energy of the system gets less. The period remains unchanged.

Fig 25
Damped oscillations

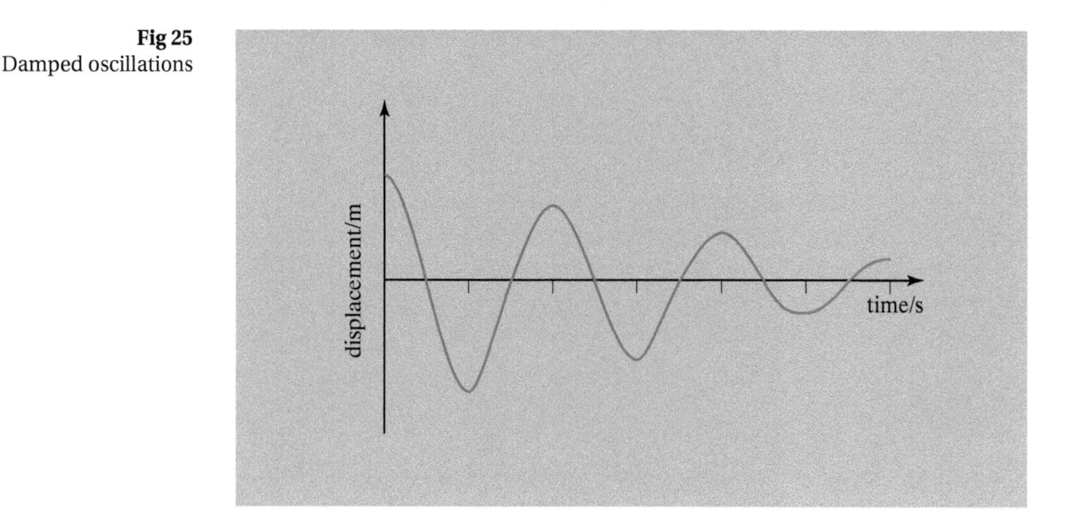

3.6.1.4 Forced vibrations and resonance

When a pendulum, a child on a swing for example, is pulled back and released, it oscillates at a steady frequency that just depends on the length of the pendulum and gravitational field strength. In this case, in the absence of any external, varying force, the pendulum undergoes **free vibrations** and oscillates at its **natural frequency**, f_0.

Definition

The natural frequency, f_0, of an oscillating system is the frequency at which the system undergoes free vibrations.

Free vibrations occur when an object is displaced from equilibrium and released. However, sometimes the vibrating object is driven by an outside force, which itself varies periodically. For example, the force you apply when pushing a child on a swing or the force from the motor of a washing machine as it causes the side panels to vibrate. These are known as **forced vibrations**. Another example is a loudspeaker, which vibrates in response to the electrical signal that drives it. These forced vibrations happen at the **driving frequency** of the electrical signals. If the driving frequency matches the natural frequency of the loudspeaker, large-amplitude vibrations will occur, which could lead to an unpleasant noise coming from the loudspeaker. This phenomenon is known as **resonance**.

Definition

Resonance occurs when a system is forced to oscillate at its natural frequency, i.e. when the driving frequency equals the natural frequency of the system.

A familiar example of resonance occurs in a children's playground when an adult is pushing a child on a swing. The adult's pushing frequency matches the natural frequency of the swing and the amplitude of the oscillations gets larger. With each push the adult is transferring energy to the swing. If this happened over many cycles the amplitude could get dangerously large. It is **damping** which prevents the amplitude of the oscillations continually increasing. Damping transfers energy from the oscillating system to the surroundings. In the case of a swing, the resistive forces of friction at the support and air resistance acting on the child will limit the size of the oscillations. In theory, if there was no damping, a resonant system would increase its energy every cycle and the amplitude would also keep increasing.

Example

Which of these oscillating systems could be classed as free vibrations and which as forced vibrations?

A a drum 'skin' vibrating after being hit with a drum stick

B an empty bus seat vibrating while the engine is idling

C a guitar string after it has been plucked once

D water molecules in a microwave oven.

Answer

A and C are free vibrations, since there is no external varying force driving the oscillations.

B and D are forced vibrations.

Resonance often causes problems in mechanical systems. In 1942 the Tacoma Narrows Bridge in the USA collapsed spectacularly when turbulence from the wind set it vibrating. The frequency of the forces from the wind eddying around the bridge matched the natural frequency of the suspension bridge and large-amplitude oscillations built up, eventually destroying the bridge. Another example of a bridge collapsing through resonance happened in Angers, France in 1850. This time the driving force came from soldiers marching across the bridge. The frequency of the marching matched a natural frequency of the bridge, resonance occurred, the bridge collapsed and 200 soldiers died.

Resonance can also be useful. In microwave cooking it is the resonant vibrations of water molecules that heat the food. The driving frequency of the electromagnetic waves matches a natural vibration frequency of water molecules and large-amplitude vibrations occur.

Fig 26
Large amplitude oscillations occur when the driving frequency of the external force matches the natural frequency of the system. Damping limits the resultant amplitude.

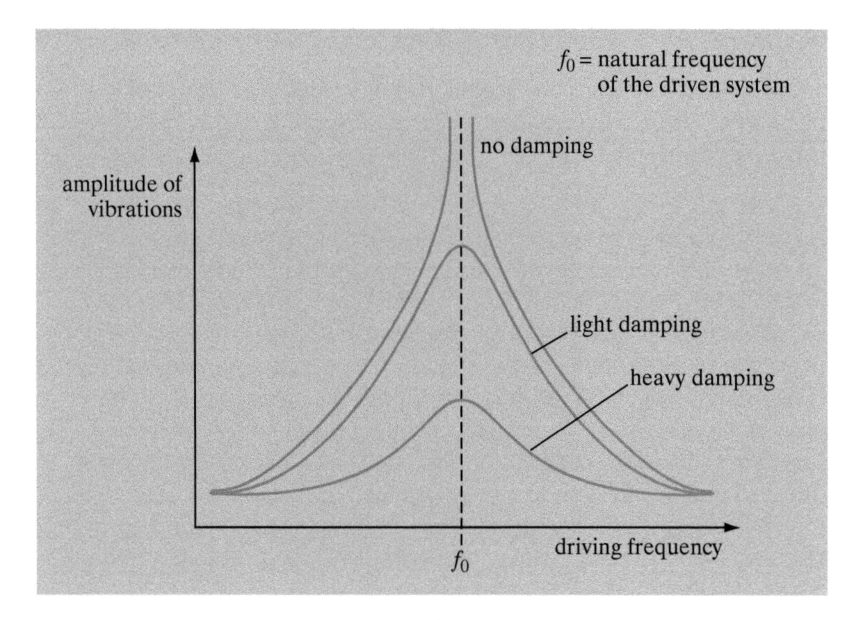

Example

The casing of a washing machine vibrates as the motor driving the drum spins round. As the motor speeds up the vibrations increase until, at a certain motor speed, the casing vibrates violently. At higher motor speeds the vibrations die away again.

(a) Explain these observations.

(b) Washing machines often have a large block of concrete bolted to the casing. Explain why.

(c) How could the amplitude of these vibrations be decreased?

Answer

(a) This is an example of resonance. When the driving frequency of the motor's oscillations matches the natural frequency of the casing, large amplitude vibrations occur.

(b) The concrete block increases the mass of the vibrating system. This lowers the natural frequency below the motor's normal running speed.

(c) Adding extra damping to the system can decrease the amplitude. A shock absorber can be fitted to the casing. The shock absorber is a piston moving in a cylinder filled with oil. As the casing vibrates, oil is forced through small holes in the piston, dissipating energy and reducing the vibrations.

Further examples of resonance

Musical instruments rely on resonance for their loudness and tone. The vibrations of the strings on an acoustic guitar provide the driving force, which makes the body of the guitar resonate. In a wind instrument, the tube of air inside the instrument is forced to resonate by the vibrations of a reed or by air blowing over a hole.

A resonating air column can be set up using a glass tube, open at both ends, but with one end in water (see Fig 27), so that the length of the air column can easily be varied. The air is forced to vibrate by a tuning fork, or a small loudspeaker. The sound from the resonance tube can be detected by ear, or by a microphone connected to an oscilloscope.

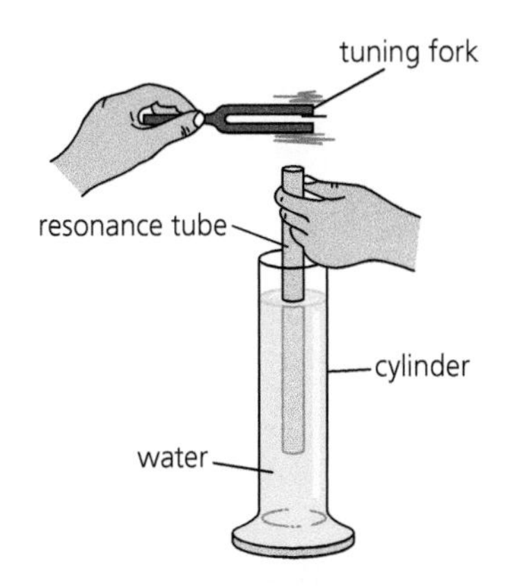

Fig 27
The resonance tube method

The air will resonate when the driving frequency (from the loudspeaker) matches the natural frequency of the air in the tube. The lowest natural (or **resonant**) frequency occurs when there is a node at the closed end of the tube and an antinode at the open end. When this happens, one quarter of a wave fits into the tube:

$$\tfrac{1}{4}\lambda = l, \qquad \text{or} \qquad 4\lambda = l.$$

Other resonances will occur at higher frequencies, $\tfrac{3}{4}\lambda = l$, $\tfrac{5}{4}\lambda = l$ and so on.

At resonance, the air in the column undergoes large amplitude oscillations, and the emitted sound is much louder than at other frequencies. Although the driving force from the loudspeaker continues, the oscillations in the tube do not continue to get larger. Energy is lost from the system as the vibrations transfer energy to the glass tube, the bench, the air outside the tube, etc. We say that the vibrations are 'damped'.

Essential Notes

Remember the work on **stationary waves** in AS/A-Level Year 1 Section 3. A node is a point of minimum vibration and an antinode is a point of maximum vibration.

Essential Notes

These are known as the harmonics. The particular mix of harmonics produced by a particular instrument determines its tone.

Example

A resonance tube is set up as in Fig 27 above, using a loudspeaker connected to a signal generator to provide the driving force. The signal generator is adjusted so that the frequency varies from 0 Hz to 1 kHz. The length of the tube is kept fixed at 20 cm. Sketch a graph showing how you would expect the amplitude of the sound waves in the tube to vary with frequency.

Answer

The first resonance (harmonic) occurs at $\lambda = 4 \times 0.2 = 0.8$ m, and as the speed of sound in air is approximately 330 m s^{-1}, the frequency will be around 410 Hz. There will be another resonance at 820 Hz.

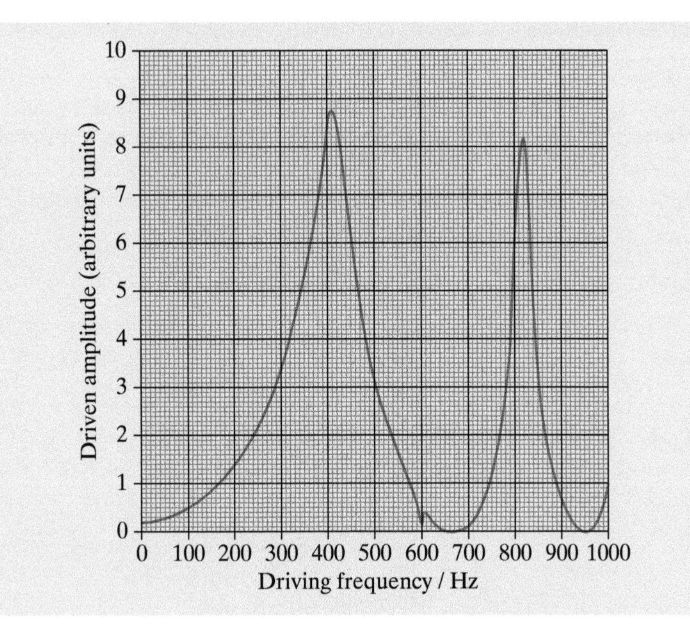

3.6.2 Thermal physics

3.6.2.1 Thermal energy transfer

Internal energy

An object may have energy due to its position, as in gravitational potential energy, or due to its motion, i.e. kinetic energy. It also has energy due to the motion and position of all the particles that it is composed of. This **internal energy** is the sum of all the kinetic energies and potential energies of its constituent particles.

In a fluid (a liquid or a gas), particles can move relative to each other and they have kinetic energy due to this motion. They may also have rotational kinetic energy. Particles in a solid oscillate about a fixed position and they have kinetic energy due to these vibrations.

Essential Notes

The internal energy of a solid or a liquid is the sum of the kinetic and potential energies of all its particles. The internal energy of a gas is due solely to the kinetic energy of its particles.

In a solid or a liquid, particles exert a force on neighbouring particles and so they have potential energy due to their position. In a gas, however, the particles are usually so far apart that these intermolecular forces can be ignored.

The internal energy of a system can be increased by heating it or by doing work on it. For example, think of the air in a bicycle pump. When the piston is pushed in, work is done on the air, making the molecules move faster and thereby raising the internal energy of the air. The internal energy of the air could also be increased by heating the pump, perhaps by dropping it into boiling water!

Fig 28
Molecular energy in a hot cup of tea

1. The molecules in the cup have KE due to their vibrational movement and PE due to the forces between them.
2. The molecules in the tea have KE due to their movement and PE due to the forces between them.
3. The molecules in the vapour have KE due to their movement, but no PE as the forces between them are negligible.

Heating an object relies on a temperature difference to transfer energy, work does not.

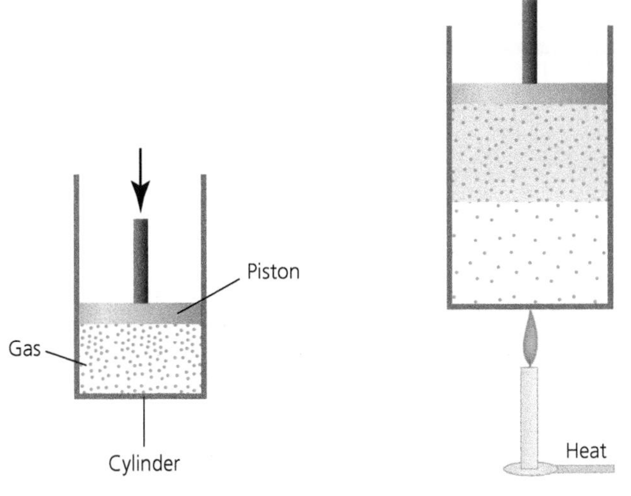

Fig 29
Heating an object

The internal energy of the gas can be raised by doing work on it. The piston is pushed down, which increases the KE of the gas molecules.

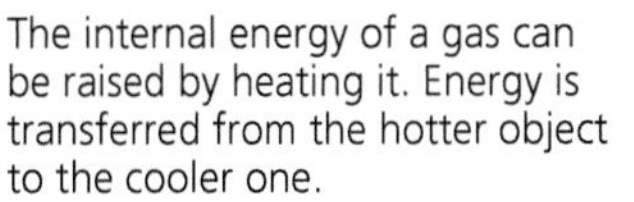

The internal energy of a gas can be raised by heating it. Energy is transferred from the hotter object to the cooler one.

Essential Notes

Heat and work are not 'things'; they are processes. They are not forms of energy; they are ways of transferring energy. When we say 'Heat flows from a hotter object into a colder one' what we really mean is 'Energy is transferred by the hot object heating the cooler one.'

Any change in the internal energy of a system, such as the air in a bicycle pump, is due to energy transfers by heating or by work. The principle of conservation of energy tells us that change in internal energy, ΔU, equals energy transfer due to work, ΔW, plus energy due to heating, ΔQ:

$$\Delta U = \Delta W + \Delta Q$$

This formulation of the conservation of energy is known as the **first law of thermodynamics**.

One effect of transferring energy to an object is to raise its temperature. This may be done by heating, for example using a kettle to bring water to the boil; or by doing work, for example a bicycle pump becomes hot as you

Essential Notes

If ΔU is negative, the internal energy of the system has been decreased. This will occur if work is done **by** the system, transferring energy to the surroundings, and/or if ΔQ is negative, indicating that the system has transferred energy by heating a cooler object.

pump up the tyres. The energy that is needed to cause a temperature rise depends on several factors:

- the mass of the system, m
- the size of the temperature rise, $\Delta\theta$
- what the system is made of.

If we fix the mass of the system, to 1 kg say, we find that different materials need different amounts of energy for the same temperature rise, say 1°C. This property of materials is known as the **specific heat capacity**, ***c***.

Definition

The specific heat capacity of a material is the energy needed to cause a temperature rise of 1 K in a mass of 1 kg. The SI unit for specific heat capacity is joules per kilogram per kelvin, J kg^{-1} K^{-1}.

'Specific' means the value per unit mass. In SI units, a unit mass is one kilogram, so specific heat capacity refers to 1 kg of a given material. Values for given materials can be looked up in data tables.

We can use the specific heat capacity to calculate the energy, ΔQ, required to heat any mass, m, of the material by any temperature rise, $\Delta\theta$.

$$\Delta Q = mc\,\Delta\theta$$

Essential Notes

Thermodynamics is literally the study of heat and motion. It deals with the transfer of energy by heat and work.

Essential Notes

A temperature change of 1 kelvin, 1 K, is identical to a temperature change of 1 °C. The Kelvin and Celsius scales have equal increments, they just start at different points.

0 K = −273.15 °C

0 °C = 273.15 K

Essential Notes

The **heat capacity** of an object, a saucepan for example, is the energy required to raise its temperature by 1 K, in J K^{-1}.

Table 2
Examples of specific heat capacities

Material	Specific heat capacity / J kg^{-1} K^{-1}
air	993
water	4190
copper	385
concrete	3350
gold	135
hydrogen	14 300

Example

A 2.2 kW electric kettle is used to heat 1.5 litres (1.5 kg) of water. Assume that all the energy is transferred to the water, ignore the energy needed to warm the kettle itself and any energy losses to the surroundings. How long will it take the kettle to bring the water, initially at 283 K, to the boil?

Answer

The total energy required is:
$\Delta Q = mc\,\Delta\theta = 1.5 \times 4190 \times (373 - 283) = 565\,650$ J

The electric kettle transfers 2200 joules every second.

$$\text{Time needed} = \frac{565\,650}{2200} = 260\,\text{s (to 2 s.f.)}$$

The specific heat capacity of water is relatively high. This means that water is useful for transferring energy, for example in water-cooled engines and in domestic central heating. The high specific heat capacity of water also means that the temperature of a large mass of water, like the sea, only changes slowly.

Example

An electric shower has a power of 7 kW. The incoming water temperature is 10 °C. If the required water temperature for the shower is 35 °C, what is the maximum flow rate that the shower can provide?

Answer

$Q / t = m / t \times c \times \Delta\theta + H$, where H is the rate of energy loss. If H can be neglected,

$7000 / (4200 \times 25) = m / t = 0.067$ kg s^{-1}. 4 kg min^{-1}, or since this is water, 4 litres per minute.

Example

In an experiment to find the specific heat capacity of brass, a 300 g brass sphere is immersed in boiling water before being quickly transferred to an insulated container holding 1 litre of water at a temperature of 20 °C. The specific heat capacity of brass is actually 377 J kg^{-1} K^{-1}.

(a) What is the highest temperature that the water in the container could reach?

(b) The largest source of uncertainty in the experiment is the measurement of temperature. Estimate the **percentage uncertainty** in the specific heat capacity due to the measurement of temperature.

(c) If you were carrying out this experiment what could you do to improve the precision of the result?

Answer

(a) The internal energy lost by the brass is transferred to the water, the container and the surroundings. If we can assume no energy is transferred to the surroundings and ignore the energy needed to warm the container itself:

Suppose the highest temperature reached by the water = θ

$0.300 \times 377 \times (100 - \theta) = 1 \times 4200 \times (\theta - 20)$

$11\,310 + 84\,000 + 400 = 113\,\theta + 4200\,\theta$

$95\,710 = 4313\,\theta$

$\theta = 22.2$ °C

(b) The **absolute uncertainty** in the temperature = ±0.1 °C. So the uncertainty in a temperature *difference* $\Delta\theta = \pm 0.2$ °C. As the temperature rise is only 2 °C this is an uncertainty in $\Delta\theta$ of 10%. Assuming the percentage uncertainty in the mass measurement is low, the percentage uncertainty in the specific heat capacity will also be 10%.

(c) To improve the precision of the result we need to use a thermometer with a higher **resolution** and/or arrange for a larger temperature increase (so that the percentage uncertainty in $\Delta\theta$ is reduced). This can be done by using a larger mass of brass and a smaller mass of water.

Essential Notes

The English scientist James Joule demonstrated that kinetic energy could be transferred to internal energy by using horses to turn paddle wheels in a tank of water. A temperature rise of the water showed that energy had been transferred.

Energy transfers

Whenever energy is transferred, some of the energy ends up increasing the internal energy of the surroundings. For example, when a cyclist freewheels down a hill, her gravitational potential energy is transferred as kinetic energy as she speeds up. But resistive forces will act to slow her down, energy will be transferred to the road, the air and the bike causing the temperature of each to increase slightly. When a hammer is used to strike metal, potential energy is transferred to kinetic energy, but the end result is that the hammer and the metal get hotter. The brakes in a car transfer kinetic energy to internal energy in the brakes, which can glow red hot.

Example

A laboratory method of demonstrating energy transfers uses lead shot in a closed cardboard tube. The tube is held vertically and turned end to end so that the lead is continually being lifted and allowed to fall, hitting the bottom of the tube. The tube is 1 m long and there is 250 g of shot in it. Lead has a specific heat capacity of 126 J kg^{-1} K^{-1}.
How much of a temperature rise would you expect in the lead after 100 inversions of the tube?

Answer

Each time the lead falls, potential energy is transferred to kinetic energy and then to internal energy. The change in potential energy = $mg\,\Delta h$ = 0.250 kg × 9.81 × 1 m = 2.45 J, so 245 J after 100 inversions.

If this is all transferred as internal energy in the lead, the temperature rise, $\Delta\theta$, will be

$$\Delta\theta = \frac{\Delta Q}{mc} = \frac{245\,\text{J}}{(0.250\,\text{kg} \times 126\,\text{J}\,\text{kg}^{-1}\,\text{K}^{-1})} = 7.8\ ^\circ\text{C}$$

In practice the temperature rise would be less than this because of heat losses to the surroundings.

Specific latent heat

When energy is transferred to an object as heat it does not always lead to an increase in temperature. The energy transfer can lead to a **change in state**, such as when ice melts or water turns to steam. The energy needed to change the state of a substance is known as its **latent heat**, ***l***. The energy required to melt 1 kg of a solid is known as the **specific latent heat of fusion**. The energy required to convert 1 kg of a liquid to a gas is known as the **specific latent heat of vaporisation**.

Definition

The specific latent heat of fusion of a substance is the energy required to change 1 kg of a solid into 1 kg of liquid, with no change in temperature.

Definition

The specific latent heat of vaporisation of a substance is the energy required to change 1 kg of a liquid into 1 kg of gas, with no change in temperature.

The energy, ΔQ, needed to change the state of a substance of mass m, is therefore:

$$\Delta Q = ml$$

Table 3
Specific latent heat values

Material	Specific latent heat of vaporisation / kJ kg^{-1}	Specific latent heat of fusion / kJ kg^{-1}
water	2260	334
oxygen	243	14
helium	25	5
mercury	290	11
iron	6339	276
lead	854	25

When a substance changes from a solid to a liquid, or from a liquid to a gas, energy is needed to do work against the attractive forces holding the solid or liquid together. However, during a change of state there is no increase in the average kinetic energy of the particles, and so there is no change in temperature. Usually there is an increase in volume as a solid melts, or a liquid boils. Energy is needed to do work against external pressure as the substance expands. The latent heat of a substance is therefore the sum of the energy needed to increase the potential energy of its particles and to do work against external pressure.

Essential Notes

When water turns to steam at atmospheric pressure (1×10^5 Pa) about 7% of the energy supplied is needed to do work against atmospheric pressure; the rest is needed to increase the potential energy of its molecules as they move apart from each other.

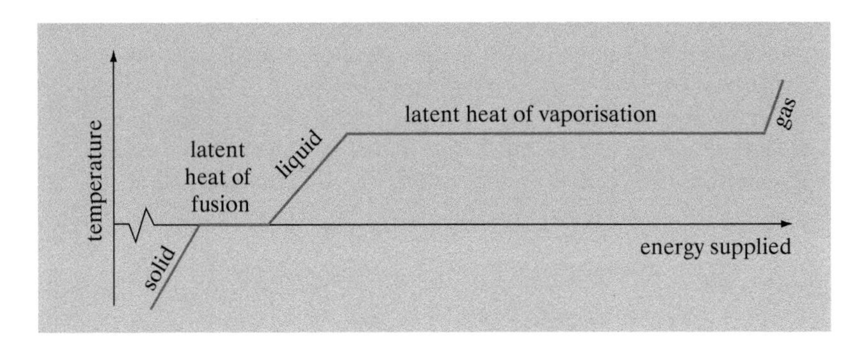

Fig 30
The temperature change of a substance as it is heated

If energy is supplied at a constant rate to a solid substance its temperature will rise until it reaches its melting point. The substance will then melt at constant temperature. When all the solid has turned into liquid, the temperature will rise again until the liquid reaches its boiling point. The substance will then boil at constant temperature until all the liquid has turned into gas, when the temperature will rise again.

The energy needed to change the state of a substance is often used to dissipate energy. The cooling towers of a power station transfer energy from the power station by evaporating large amounts of water.
We do the same thing on a smaller scale – when we sweat
we are transferring excess internal energy by evaporating liquid.

Example

An ice cube of mass 20 g is added to 200 g of water which is initially at 20 °C. As the ice melts it has a cooling effect on the water. If all the energy needed to melt the ice comes from the water, calculate the final temperature of the water.

Answer

The latent heat necessary to melt the ice is

$$\Delta Q = ml = 0.020\,\text{kg} \times 334 \times 10^3\ \text{J kg}^{-1} = 6680\,\text{J}$$

If this energy all comes from the water:

$$\frac{\Delta Q}{mc} = \Delta T = \frac{6680\,\text{J}}{(0.200\,\text{kg} \times 4190\,\text{J kg}^{-1}\,\text{K}^{-1})}$$

$$\Delta T = 8.0\ ^\circ\text{C}$$

The temperature of the water will drop to around 12 °C because of the ice melting.

Note however that we have neglected here the energy needed to raise the temperature of the 20 g of water from the melted ice at 0 °C to the final temperature of the water. If this also comes only from the water, the final temperature will be about 11 °C.

Example

The amount of water that we lose through the evaporation of sweat depends on the temperature of our surroundings, as well as on the humidity and wind speed. On average a typical person loses about 0.5 litre of sweat per day. Calculate the average power of this energy transfer.

Answer

The energy needed to evaporate 0.5 litre of sweat is $\Delta Q = ml$.

1 litre of water has a mass of 1 kg, so $m = 0.5$ kg.

l is the specific latent heat of vaporisation of water = $2.260 \times 10^6\ \text{J kg}^{-1}$.

So the energy transferred, $\Delta Q = 0.5\ \text{kg} \times 2.260 \times 10^6\ \text{J kg}^{-1} = 1.13 \times 10^6\ \text{J}$.

If this is transferred over a period of 24 hours or $24 \times 60 \times 60 = 86\,400$ seconds, the average power is:

$$\frac{1.13 \times 10^6}{86400} = 13\,\text{W}$$

3.6.2.2 Ideal gases

The physical state of a fixed mass of gas can be described by three quantities: pressure, temperature and volume.

Pressure The pressure, p, that a gas exerts on the walls of its container is caused by the collisions of the molecules with the walls. Pressure is defined as the force per unit area and is measured in pascals, Pa. 1 Pa is a pressure of 1 newton per square metre, $1 \text{ Pa} = 1 \text{ N m}^{-2}$.

Temperature The temperature, T, of the gas is a measure of the average kinetic energy of its molecules. T is the absolute temperature measured in degrees kelvin, K.

Volume The volume, V, is the space occupied by the gas, and is measured in m^3.

Essential Notes

The resistance of a metal wire, the peak wavelength of radiation emitted by an object, the volume of a gas and the e.m.f. produced at the junction of two different metals are all thermometric properties and are used in different types of thermometers.

Temperature scales

The temperature of a gas is proportional to the mean kinetic energy of its molecules. This cannot be measured directly. Instead, the temperature is measured by its effect on another property, such as the length of a column of mercury in a glass tube or the resistance of a wire. These are known as **thermometric properties**.

Different thermometric properties vary with temperature in different ways. This could mean that temperature readings taken in the same place, at the same time, would be different. To avoid this, we define a standard scale known as the **absolute temperature scale**, also known as the **Kelvin scale**. This scale is based on the pressure of an ideal gas (see Essential Note).

As an ideal gas is cooled, its molecules slow down and the gas exerts a lower pressure on the walls of its container. If the container has a fixed volume and holds a fixed mass of gas, the pressure, p, of the gas is proportional to the absolute temperature, T, as shown in Fig 31. By extrapolating the best-fit line backwards until it meets the x-axis, we can find the temperature at which, theoretically, the gas will exert no pressure. This temperature is known as **absolute zero**, 0 kelvin or 0 K. This is the lowest conceivable temperature; the molecules have no kinetic energy.

Essential Notes

The forces between gas molecules are relatively small and the molecules are much further apart than those in a liquid or solid. An ideal gas is one in which the forces are equal to zero. A dry gas at low pressure is a close approximation to an ideal gas.

Essential Notes

In practice, gases may liquefy as they are cooled. According to quantum theory, molecules have some energy, known as zero point energy, at 0 K.

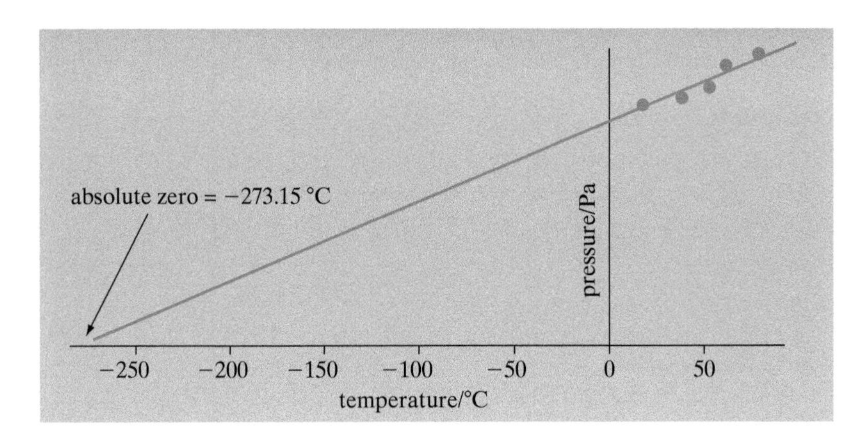

Fig 31
Pressure vs temperature for an ideal gas

Essential Notes

It is theoretically impossible to cool something to absolute zero, but researchers have got very close. Liquid helium has been cooled to 90 μK (1 μK is 1 microkelvin or 10^{-6} K).

A graph plotted of pressure against temperature can be extrapolated backwards to very low temperatures. Eventually the pressure of an ideal gas would drop to zero. At this point the molecules have stopped moving and the gas cannot get any colder. This point is the lowest conceivable temperature and it is known as **absolute zero**.

Definition

Absolute zero, 0 K, is equal to −273.15°C. So to convert a temperature in degrees celsius to kelvin you need to add 273.15.

The gas laws

A gas can be described in terms of its bulk (macroscopic) properties. These are:

- mass, m
- pressure, p
- volume, V
- temperature, T.

Because the molecules in an ideal gas do not interact with each other, these properties are linked by simple relationships, known as the gas laws. These were first found experimentally in the 1600s.

1. Boyle's law

The volume of a fixed mass of gas can be changed by altering the pressure on it. If you block the end of a bicycle pump with your finger and push the piston in, you increase the pressure on the air and its volume decreases. In fact, for a fixed mass of an ideal gas at constant temperature the volume, V, is inversely proportional to the pressure.

Definition

For a fixed mass of an ideal gas at constant temperature $p \propto \frac{1}{V}$

For a fixed mass of gas at constant temperature therefore

$$pV = \text{constant} \quad \text{or} \quad p_1V_1 = p_2V_2$$

Notes

Boyle's law only applies to a fixed mass of gas at constant temperature. You couldn't apply $P_1V_1 = P_2V_2$ to a tyre or a balloon that was being inflated since the new pressure and volume are affected by the increased number of gas molecules.

Example

20 cm^3 of air at atmospheric pressure, 1×10^5 Pa, is trapped in a bicycle pump when a finger is placed over the end of the pump. If the piston is pushed in until the volume is 5 cm^3, find the new pressure.

Answer

$p_1V_1 = p_2V_2$

$20\ \text{cm}^3 \times 1 \times 10^5\ \text{Pa} = 5\ \text{cm}^3 \times p_2$

$p_2 = 4 \times 10^5\ \text{Pa}$

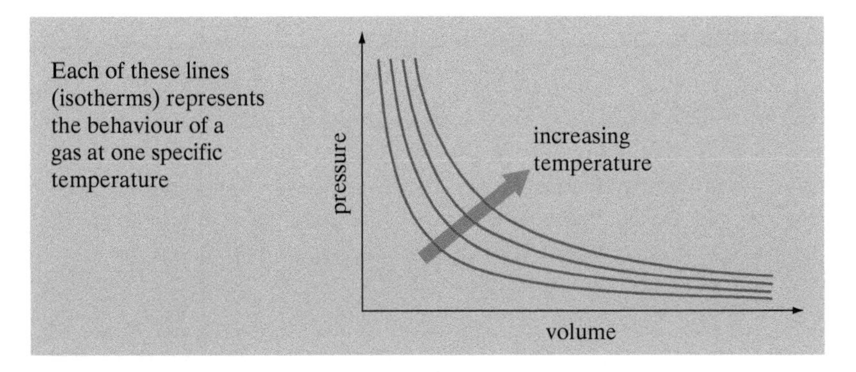

Fig 32
The lines on the graph all show that pressure is inversely proportional to volume. Each of the lines, which are known as isotherms, represents a different temperature.

2. Charles' law

When gases are heated at constant pressure, they all expand at the same rate. The volume of a gas is proportional to its temperature, providing we use the Kelvin temperature scale.

Fig 33
V vs T for an ideal gas at constant pressure

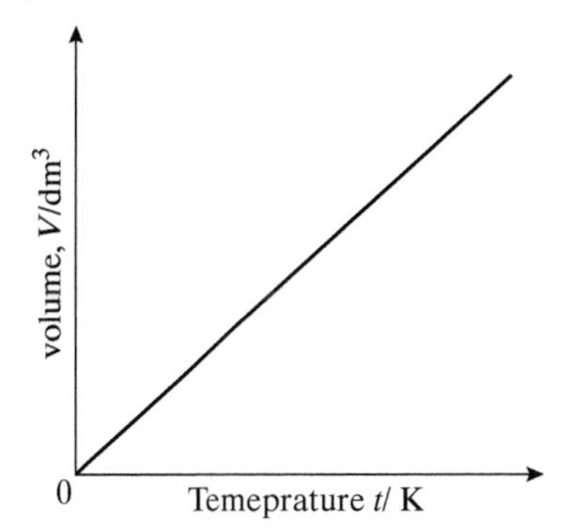

Definitions

The volume of a fixed mass of an ideal gas at constant pressure is proportional to its absolute temperature, V ∝ T.

3. The pressure–temperature law

If a gas is heated in a container of fixed volume its pressure will increase. In fact the pressure of a gas is proportional to its absolute temperature.

Definitions

The pressure of a fixed mass of an ideal gas at constant volume is proportional to its absolute temperature, P ∝ T.

The equation of state

For a fixed mass of gas, the three gas laws can be summarised as:

$p \propto T$ at constant volume

$V \propto T$ at constant pressure

$p \propto \frac{1}{V}$ at constant temperature

It is possible to combine these three laws into an **equation of state for an ideal gas**:

$$pV \propto T \text{ or } pV = RT, \text{ where } R \text{ is a constant.}$$

The value of the constant, R, depends on the number of molecules in the gas. In fact, we choose the value of R to be correct for 1 **mole** of gas molecules. 1 mole is the SI unit of amount of substance; 1 mole of gas contains 6.02×10^{23} molecules (see next section).

$$pV = nRT$$

Definition

*The **equation of state for an ideal gas**, or **ideal gas equation**, is pV = nRT.*

In many problems the mass of gas is fixed, such as in the cylinder of an engine or in a sealed balloon or car tyre. The initial values of the gas's pressure, volume temperature can be written, p_1, V_1 and T_1 and the equation of state can be written:

$$\frac{p_1V_1}{T_1} = nR$$

If the gas is compressed or heated then its pressure, volume and temperature will change to new values, p_2, V_2 and T_2, but n remains fixed. The equation of state is now:

$$\frac{p_2V_2}{T_2} = nR$$

This gives a useful form of the equation:

$$\frac{p_1V_1}{T_1} = \frac{p_2V_2}{T_2}$$

Example

A bubble of air escapes from a diver's breathing apparatus at a depth of 45 m. The bubble has a volume of 2.0×10^{-5} m^3. The water pressure at a depth of 45 m is 450 kPa and the water temperature is 5 °C. What is the volume of the bubble when it has risen to the surface, where the temperature is 10 °C? Take atmospheric pressure as 100 kPa.

Answer

The total pressure at a depth of 45 m is 550 kPa, 450 kPa due to the water and 100 kPa due to atmospheric pressure.

The initial conditions of the gas are: p_1 = 550 kPa; $V_1 = 2.0 \times 10^{-5}$ m^3; T_1 = 278 K

The final conditions of the gas are p_2 = 100 kPa; V_2 is unknown and T_2 = 283 K

The mass of air in the bubble is fixed so

$$\frac{p_1V_1}{T_1} = \frac{p_2V_2}{T_2}$$

$$V_2 = \frac{p_1V_1T_2}{p_2T_1} = \frac{550\,\text{kPa} \times 2.0 \times 10^{-5}\,\text{m}^3 \times 283\,\text{K}}{100\,\text{kPa} \times 278\,\text{K}}$$

$$V_2 = 1.1 \times 10^{-4}\,\text{m}^3$$

Notes

Don't forget that you must use the absolute temperature, i.e. the temperature must be in degrees kelvin, not Celsius. Add 273 to the Celsius value to convert to kelvin.

The Avogadro, Boltzmann and molar gas constants

The equation of state for a fixed mass of an ideal gas, $pV = nRT$, can be rewritten as:

$$n = \frac{pV}{RT}$$

The number of moles of gas, n, depends only on the pressure, volume and temperature. **Avogadro's law** puts this another way:

'At the same temperature and pressure, equal volumes of gas contain equal numbers of molecules.'

If we specify a standard temperature and pressure (STP), a mole of any gas will occupy the same volume.

- Standard temperature: 0 °C = 273.15 K
- Standard pressure = 1 atmosphere = 101.3 kPa.
- Standard volume of 1 mole of an ideal gas at STP = 22.4 l or 22.4×10^{-6} m^3.

The mole can refer to any particles: electrons, ions, atoms, molecules etc. The number of particles in a mole is 6.02×10^{23}, which is known as the **Avogadro constant**.

Definition

The Avogadro constant, N_A, is the number of particles in a mole of substance.

$N_A = 6.02 \times 10^{23}$

The **molar mass**, that is the mass of 1 mole of a substance, is its relative molecular mass expressed in grams. A mole of hydrogen therefore has a mass of only 1 g whilst a mole of carbon-12 has a mass of 12 g.

Note that the **molecular mass** is linked to the molar mass by the Avogadro constant:

$$\text{molar mass} = \text{molecular mass} \times \text{Avogadro constant}$$

so that the molecular mass of carbon-12 is:

$$\frac{\text{molar mass}}{N_A} = \frac{12}{6.02 \times 10^{23}} = 1.99 \times 10^{-23}\ \text{g or } 1.99 \times 10^{-26}\ \text{kg}$$

Notes

Note that molar masses and molecular masses are sometimes expressed in grams, rather than kilograms. Relative molecular mass has no unit.

We can relate the pressure, volume and temperature of a gas to the *number of molecules* in the sample of gas, N, rather than the number of moles, n. Since $pV = nRT$, and $n = N/N_A$, the ideal gas equation can be written:

$$pV = \frac{NRT}{N_A}$$

The **Boltzmann constant**, k, is defined as the molar gas constant divided by the Avogadro constant.

Definition

The Boltzmann constant, k, is the molar gas constant, R, divided by the Avogadro constant, N_A.

$$k = \frac{R}{N_A}$$

k has a value of $1.38 \times 10^{-23} J K^{-1}$.

So, for N molecules of gas, the ideal gas equation can be written:

$$pV = NkT$$

Example

A car tyre has a volume of around 1.50×10^{-2} m^3 and it has been inflated to a pressure that is twice atmospheric pressure. If the temperature is 20 °C, estimate the mass of air in the tyre. (The relative molecular mass of air is about 29; atmospheric pressure is 100 kPa.)

Answer

$$\frac{pV}{RT} = n = \frac{200 \times 10^3 \text{Pa} \times 1.50 \times 10^{-2} \text{m}^3}{8.31 \text{J mol}^{-1} \text{K}^{-1} \times 293 \text{K}} = 1.23 \text{mole}$$

A mole of air would have a mass of 29 g so the mass of air in the tyre is about $1.23 \times 29 = 36$ g.

Work done by an expanding gas

When a gas expands, it has to push back its surroundings. The gas does external work, W. Consider the gas in the cylinder in Fig 34. If the gas pressure, P, is constant and it pushes the piston through a distance, Δs, then:

work done = force × distance moved

$$W = F \times \Delta s$$

But as pressure = $\frac{\text{force}}{\text{area}}$, $F = PA$, where A is the cross-sectional area of the cylinder.

So $W = PA\Delta s$; but $A\Delta s$ is the change in volume, ΔV, so

$$W = p\Delta V = p(V_f - V_i)$$

where ΔV is the small increase in volume of the gas. Note that $p\Delta V$ is negative if the gas expands: this means that work is being done *by* the gas to push out the piston (see Fig 34). If the piston is used to compress the gas, work is done *on* the gas and so $p\Delta V$ is positive.

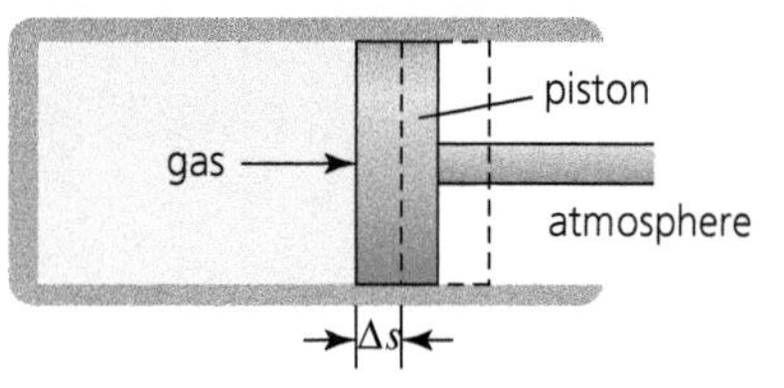

Fig 34
A gas can do work when it expands. The force exerted by the gas on the piston is pA, where p is the gas pressure in the cylinder and A is the cross-sectional area of the piston. The volume change $\Delta V = A\,\Delta s$, so the work W done by a gas expanding at constant pressure is $W = p\,\Delta V$.

Molecular kinetic theory

The gas laws discussed in the previous section describe the behaviour of a gas as a whole. They show how the bulk properties of a gas, such as pressure, volume and temperature, are linked together. These laws are empirical, which means that they are based on experimental results, but they do not explain why gases behave as they do.

The kinetic theory of gases is an attempt to explain these laws in terms of the microscopic behaviour of gas molecules. The first direct evidence that gases (and liquids) were composed of small, randomly moving particles came in 1827 when Robert Brown observed pollen through a microscope. He noticed that the particles of pollen, which were suspended in water, were in constant random motion. He concluded that the pollen was being bombarded on all sides by randomly moving molecules of water, which were themselves too small to be seen.

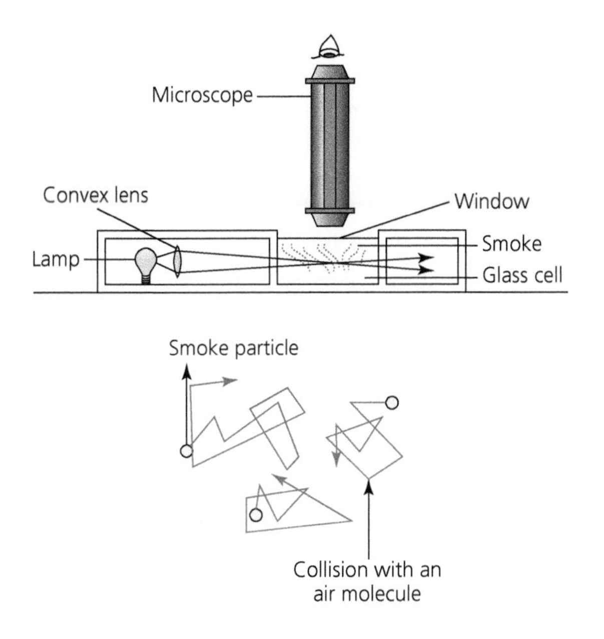

Fig 35
Brownian motion can be observed in the school laboratory using this apparatus. Small particles of smoke are seen to move randomly.

Brownian motion suggests that a gas is composed of molecules which are in constant motion. The molecules move in random directions with a range of speeds. The properties of a gas, such as temperature and pressure are due to these moving molecules. Temperature depends on the average kinetic energy of the molecules. Pressure arises due to the molecules colliding with an object, the walls of a container for example. Using kinetic theory we can derive the equations that link bulk properties, such as temperature and pressure, with microscopic properties, such as the velocity of the molecules.

Pressure of an ideal gas

An ideal gas obeys the equation $pV = nRT$ exactly. However, real gases deviate from this law, especially at high pressures and low temperatures. This is because under these conditions the gas molecules are close enough to exert a force on each other. A theoretical ideal gas would have the following properties:

- Its molecules are so small compared to the volume of the gas as a whole, that the volume occupied by the molecules themselves can be neglected.
- All collisions involving the molecules are elastic. There is no loss of kinetic energy in these collisions.
- The time taken by a collision is much less than the time between collisions.
- The forces between molecules are small enough to be ignored.

The molecules move in random directions.

Using these assumptions, we can develop an expression for the pressure of a gas in terms of the speed of its molecules.

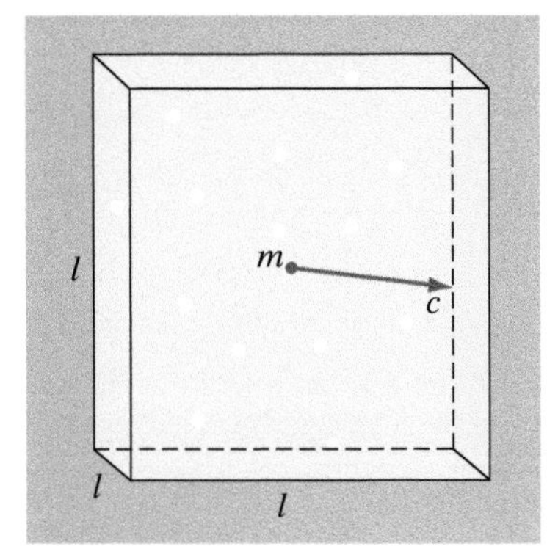

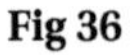
Fig 36

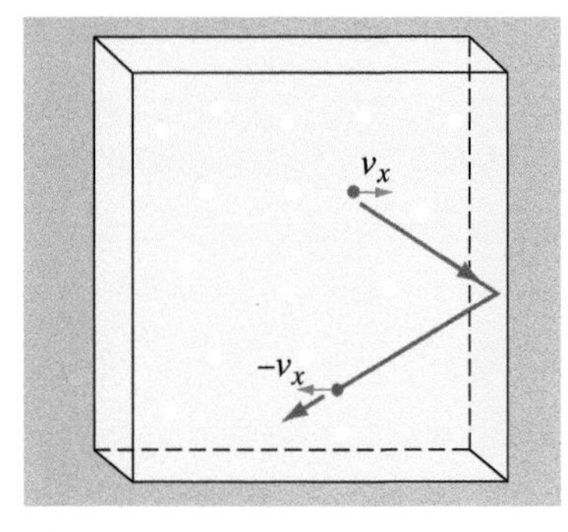

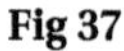
Fig 37

We start by imagining a single molecule, along in a cubical box with sides of length l metres (Fig 36). The molecule has a mass, m, and is moving in a random direction with velocity, c. Initially, we will consider only the component of the molecule's velocity in the x direction, v_x. We will choose this direction to be perpendicular to the end wall of the box. When the molecule collides with the wall, it will bounce back at the same speed as it approached, but in the opposite direction. The component of the velocity in the x direction will then be $-v_x$.

The wall exerts a force on the molecule equal to the rate of change of the molecule's **momentum**:

$$F = \frac{\text{change in momentum}}{\text{time taken}}$$

The change in momentum = final momentum – initial momentum

$$= -mv_x - mv_x = -2mv_x$$

The time, t, between collisions is the time it takes the molecule to travel across the box and back.

$$t = \frac{2l}{v_x}.$$

Force is defined by Newton's second law as the rate of change of momentum, so the force, F, exerted on the molecule is:

$$F = \frac{\text{change in momentum}}{\text{time taken}} = \frac{-2mv_x}{2l/v_x} = -\frac{mv_x^2}{l}$$

By Newton's third law, the force exerted *on the wall* is equal to this but in the opposite direction. Pressure is force per unit area. The area of one wall is l^2, so

$$p = \frac{mv_x^2}{l^3}$$

Since l^3 is the volume, V, of the box, this equation becomes:

$$p = \frac{mv_x^2}{V}$$

The total pressure is the pressure from all of the other particles as well:

$$p = p_1 + p_2 + p_3 + \ldots = \frac{mv_{x1}^2}{V} + \frac{mv_{x2}^2}{V} + \frac{mv_{x3}^2}{V} + \ldots$$

$$= \frac{m(v_{x1}^2 + v_{x2}^2 + v_{x3}^2 + \ldots)}{V}$$

The sum in the brackets is the total of all the molecules' squared velocities. This is equal to the mean squared speed, $\overline{v_x^2}$, multiplied by the total number of molecules, N, in the box. So the equation becomes:

$$p = \frac{N m \overline{v_x^2}}{V}$$

Essential Notes

The mean *velocity* of all the molecules is always zero, because they move randomly in all directions.

This is the pressure on a wall due to the x-component of the velocity of all the molecules in the box. Finally we must take into account the other components of velocity.

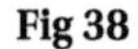

Fig 38

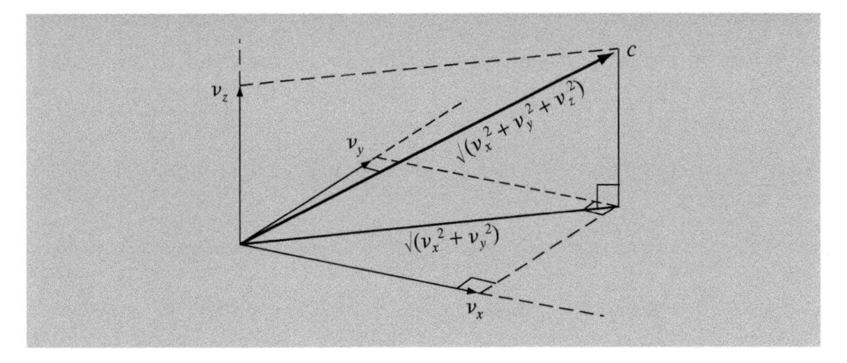

The speed of the particle c is linked to its components in the x, y and z directions by Pythagoras' theorem (see Fig 38):

$$c^2 = v_x^2 + v_y^2 + v_z^2$$

Because the motion is random, the molecules are equally likely to be moving in any of the three directions so the mean value of v_x^2, v_y^2 and v_z^2 will be the same: $\overline{c^2} = 3\overline{v_x^2}$.

So we now have this expression for the pressure:

$$p = \frac{N m \overline{c^2}}{3V} \quad \text{or} \quad pV = \frac{1}{3} N m \overline{c^2}$$

This equation links the pressure and volume of a gas to the number of molecules and their mean squared speed.

If we define a quantity c_{rms} as the **root-mean-square (r.m.s.) speed** of the molecules, that is

$$c_{rms} = \sqrt{\frac{(c_1^2 + c_2^2 + c_3^2 + ...)}{N}}$$

then

$$c_{rms}{}^2 = \frac{(c_1^2 + c_2^2 + c_3^2 + ...)}{N} = \overline{c^2}$$

and we can write the above equation as

$$pV = \frac{1}{3} Nmc_{rms}{}^2$$

Notes

You need to recall how to derive the expression for pV given here.

Since Nm is the total mass of the gas, Nm/V is actually the density, ρ, of the gas. The equation can be written:

$$p = \frac{1}{3}\rho\overline{c^2} = \frac{1}{3}\rho c_{rms}{}^2$$

Internal energy: relation between temperature and average molecular kinetic energy

As described in 3.6.2.1, the **internal energy** of a substance is the sum of the potential and kinetic energies of all its particles. In an ideal gas the particles are so far apart that intermolecular forces can be disregarded. The particles have no potential energy. The internal energy of a gas is therefore entirely due to the kinetic energy of the molecules. At any given time the total kinetic energy is shared randomly between all the molecules in the gas.

The range of molecular speeds on a gas depends on the temperature of the gas, as shown in Fig 39.

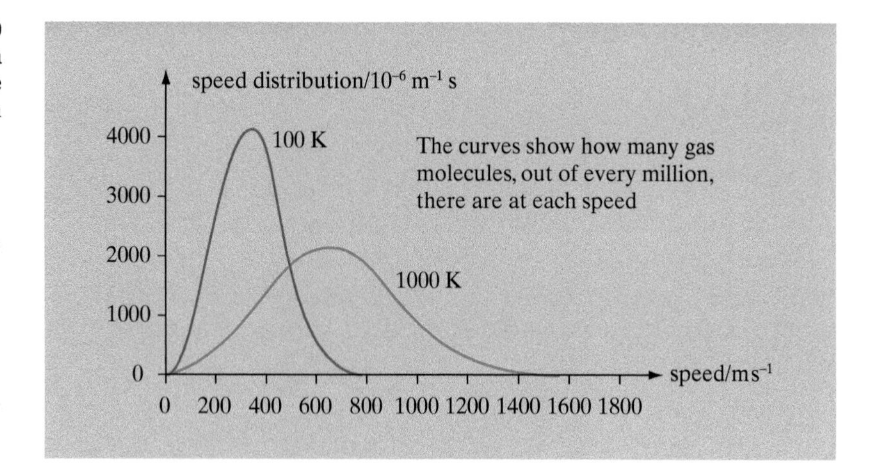

Fig 39
The range of molecular speeds at a given temperature is given by the Maxwell-Boltzmann distribution

Essential Notes

You need to know that, although the actual speed of a given molecule is random within the range, the most probable speed increases at higher temperatures.

For 1 mole of gas, kinetic theory leads to the equation:

$$pV = \frac{1}{3} N_A m c_{rms}^2$$

Compare this with the ideal gas equation:

$$pV = RT$$

These two expressions are equivalent and so:

$$\frac{1}{3} N_A m c_{rms}^2 = RT$$

This equation then lets us link the temperature of a gas to the average kinetic energy of its molecules, which is $\frac{1}{2} m c_{rms}^2$:

$$\frac{1}{2} m c_{rms}^2 = \frac{3}{2}\left(\frac{R}{N_A}\right) T$$

The ratio R/N_A is the Boltzmann constant, k (see page 42).

So the equation can be written:

$$\frac{1}{2} m c_{rms}^2 = \frac{3}{2} kT$$

Example

Calculate the average kinetic energy of the molecules of a gas at 20 °C. If the gas is oxygen, find the r.m.s. speed of the molecules. The relative molecular mass of oxygen is 32.

Answer

$$\frac{1}{2} m c_{rms}^2 = \frac{3}{2} kT = 1.5 \times 1.38 \times 10^{-23}\,\text{J K}^{-1} \times 293\,\text{K} = 6.07 \times 10^{-21}\,\text{J}$$

If the mass of 1 mole of oxygen is 32×10^{-3} kg, then the mass of 1 molecule =

$$\frac{32 \times 10^{-3}}{N_A} = 5.32 \times 10^{-26}\,\text{kg}$$

This gives a value for the mean squared speed of:

$$c_{rms}^2 = \frac{2 \times 6.07 \times 10^{-21}\,\text{J}}{5.32 \times 10^{-26}\,\text{kg}} = 2.28 \times 10^5\,\text{m}^2\text{s}^{-2}$$

Therefore the r.m.s. speed of an oxygen molecule at 20 °C is about 480 m s^{-1}.

3.7 Fields and their consequences

3.7.1 Fields

Essential Notes

If you study Physics beyond A-level you will need to use a wider definition of a field. At A-level you should know that the fields you are studying are vector fields. Each point in the field has a vector associated with it that determines the magnitude and direction of the force on an object placed there.

Fields in physics

The solar system is a collection of objects: planets, comets and other bits of rock, which orbit the Sun. These objects are kept in orbit by the attractive force of the Sun's gravity. The force is stronger closer to the Sun, and weaker further away. A mass placed at any point in this region will be subject to a force; the size and direction of which depends on its position relative to the Sun. We refer to this region of space as the **gravitational field** of the Sun.

This idea of a field as a region of space where an object is subject to a force is useful throughout physics. At A-level you study:

- the gravitational field due to a spherical mass
- the electric field due to a charge
- the magnetic field due to a moving charge.

3.7.2 Gravitational fields

3.7.2.1 Newton's law

Gravity is the attractive force that acts between all masses. Isaac Newton realised that gravity was a universal force that not only made objects fall towards the Earth's surface, but also kept the Earth and planets in their orbits around the Sun. In his law of gravitation Newton suggested that the attractive force, F, between two masses, m_1 and m_2 is:

- proportional to the product of the masses, $F \propto m_1 m_2$;
- inversely proportional to the square of the distance between the masses, $F \propto 1/r^2$.

Notes

This equation is given on the data sheet. Make sure you know what each symbol represents.

These rules combine to form **Newton's law of gravitation:**

$$F = \frac{G m_1 m_2}{r^2}$$

where G is the universal gravitational constant.

Newton could not measure the size of G and so was unable to confirm this equation directly. Instead he tested his hypothesis by using it to predict the orbital period of the Moon. The gravitational constant was later measured and found to be 6.67×10^{-11} N m^2 kg^{-2}. The reason why Newton could not measure G directly, and the reason why G is still known with less precision than many other constants, is that the gravitational force is quite weak. The force between two identical 1 kg masses placed 1 m apart is equal to:

$$F = \frac{Gm_1m_2}{r^2} = \frac{6.67 \times 10^{-11} \times 1 \times 1}{1^2} = 6.67 \times 10^{-11}\,\text{N}$$

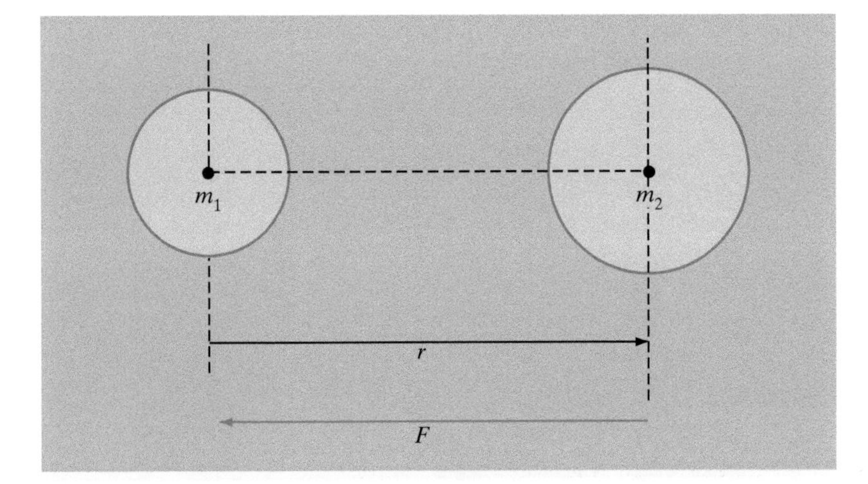

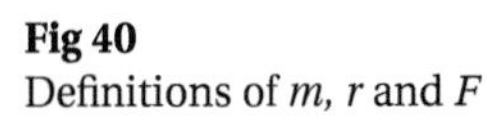
Fig 40
Definitions of m, r and F

Example

Estimate the attractive gravitational force between two students sitting next to each other in a classroom. How long would it be before this force acting alone would lead to a noticeable speed towards each other?

Answer

Estimate the mass of the students as 60 kg each, and the distance between them (between their centres of mass) as 1 m. If we apply Newton's law of gravitation:

$$F = \frac{Gm_1m_2}{r^2} = \frac{6.67 \times 10^{-11} \times 60 \times 60}{1^2} = 2.4 \times 10^{-7}\,\text{N}$$

This force acting on its own would cause one of the students to accelerate towards the other at around 4×10^{-9} m s^{-2}. It would take about 8 years to reach a speed of 1m s^{-1} at this rate! (Use $v = u + at$) No wonder you don't feel the attraction!

3.7.2.2 Gravitational field strength

The **gravitational field strength**, g, at a point is defined as the force that acts on a unit mass placed at that point. It is a vector quantity.

Definition

Gravitational field strength at a point is the force per unit mass, $g = F/m$.

The SI unit of gravitational field strength is N kg^{-1}.

The force due to a gravitational field can be represented by **field lines.** The field lines are marked with an arrow to show the direction of the force that would act on a mass placed there. Near the Earth's surface the field lines are approximately parallel to each other and perpendicular to the surface of the Earth. The field here can be treated as uniform and the strength doesn't change significantly with height. On a larger scale the field lines around a spherical mass spread out radially.

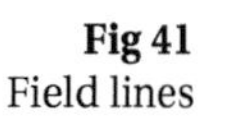

Fig 41
Field lines

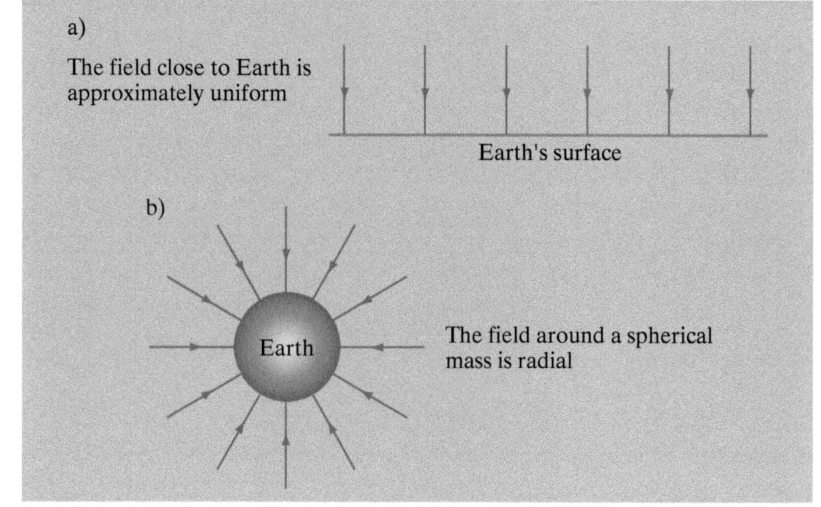

Essential Notes

Gravitational field strength is a vector quantity. Its direction is that of the force acting on a mass in the field.

For a radial field, such as that due to a spherical mass M, the force, F, that acts on a small mass, m, placed at a distance, r, is given by Newton's law as $F = GMm/r^2$. The field strength is $g = F/m = GM/r^2$.

Definition

For a radial gravitational field, the magnitude of the field strength at a distance r from a mass M is $g = GM/r^2$.

Fig 42
Gravitational field strength versus distance (radial field)

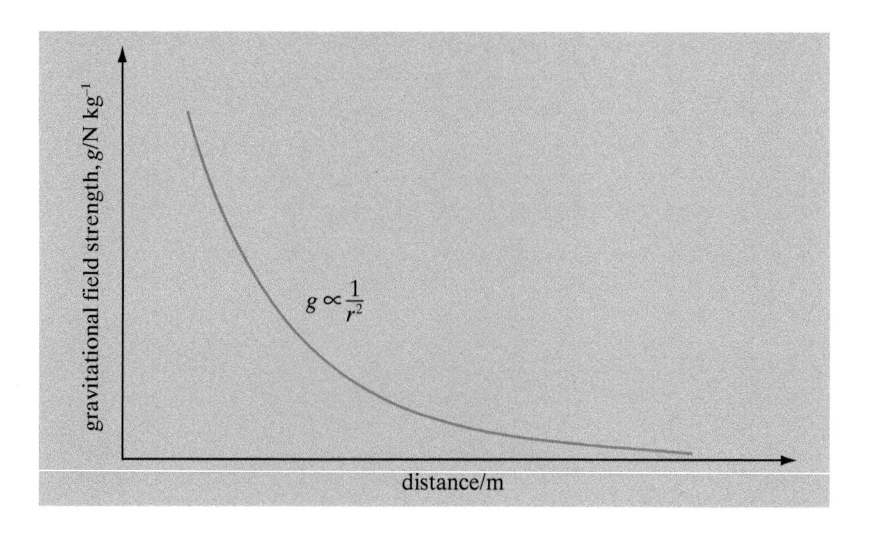

3.7.2.3 Gravitational potential

An object in the Earth's gravitational field, for example a ball at the top of a hill, has potential energy due to its position. As the ball rolls down the hill, some of this gravitational potential energy will be transferred as kinetic energy. If the ball has a mass m and falls through a height Δh, the change in its gravitational potential energy, ΔE_p, will be:

$$\Delta E_p = mg\Delta h$$

If the ball is then raised through a height Δh, the gain in gravitational potential energy will be equal to the work done in lifting it.

$$\Delta E_p = mg\Delta h = \text{force} \times \text{distance} = (mg) \times \Delta h$$

This is approximately true for small changes in the height because the gravitational field strength, g, is constant over this distance.

However, for larger values of Δh we need to take into account that the gravitational field strength is not constant (see Fig 43).

Suppose we move a mass a small distance Δr away from the Earth.

Since work done = force × distance moved, and the points are a distance Δr apart, then the work done, ΔW, is:

$$\Delta W = (GMm/\ r^2) \times \Delta r$$

provided that Δr is small enough so that the force can be treated as constant over that distance.

This is the area of the strip shaded purple in Fig 43.

The total work done, W, in removing the mass, m, from the gravitational field is the sum of all such shaded strips. This is the total area below the curve (the area shaded pink in Fig 43) which is given by the equation :

$$W = GMm/r$$

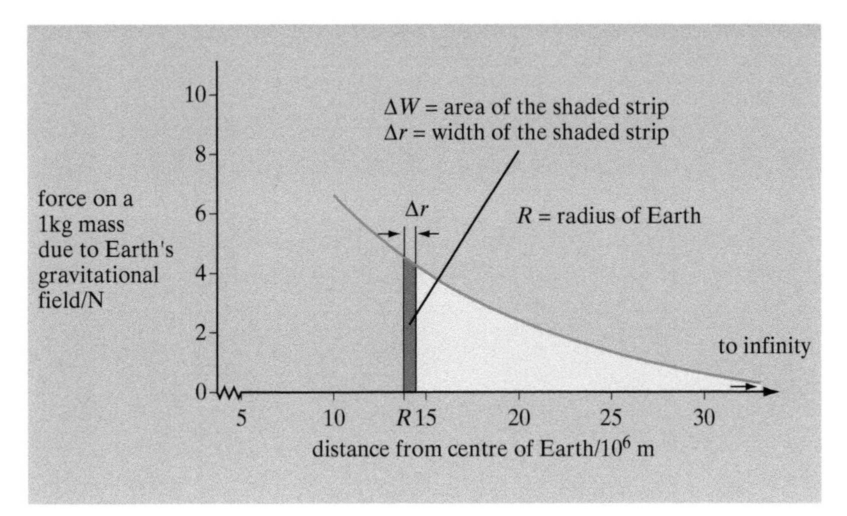

Fig 43
Work done in moving 1 kg mass from infinity to a distance r from Earth

Essential Notes

Gravitational potential is a scalar quantity; it has no associated direction.

The work done in moving a mass between any two points in a gravitational field is the area below the curve between these two points (see Fig 44).

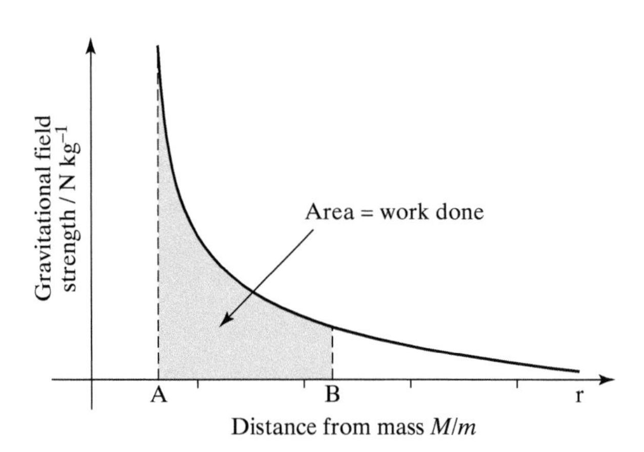

Fig 44
The work done in moving a unit mass from point A to point B is equal to the area below the curve.

So the gravitational potential, V, around a spherical mass, like the Sun or a planet, is inversely proportional to the distance from the centre of the mass

$V \propto -1/r$

V is equal to zero at infinity, and gets more negative nearer the mass.

The value of the gravitational potential at a point is telling you how much energy is needed to move a unit mass (in SI units that is 1 kg) from that point to infinity.

We define a quantity called gravitational potential, V, to help us to calculate energy changes in gravitational fields.

Essential Notes

The negative sign arises because V is defined as the work done in bringing the mass **from** infinity. As the gravitational force is attractive, positive work has to be done on the mass to move it **to** infinity.

Definition

The gravitational potential, V, at a point in a gravitational field is defined as the work done in bringing a unit mass from infinity to that point.

Gravitational potential is defined for a unit mass, so m = 1 kg, and

$$V = -GM/r$$

Notes

Notice that the gravitational potential is proportional to $1/r$, whilst the gravitational field strength is proportional to $1/r^2$. If you are asked to sketch these graphs in an exam, try to show field strength falling away more steeply with distance than the potential does.

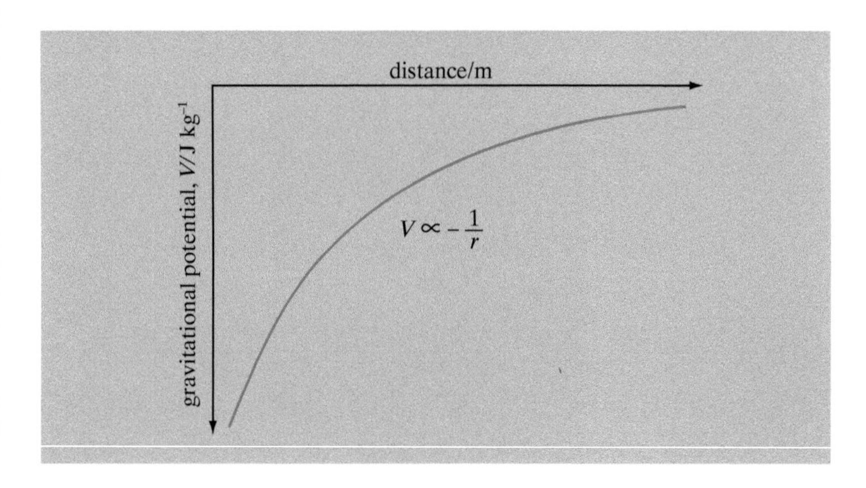

Fig 45
The variation of gravitational potential with distance (radial field)

Gravitational potential is the work done (or energy required) per kilogram. It has units of joules per kilogram, J kg^{-1}. To convert potential to potential energy, you need to multiply by the actual mass involved:

change in gravitational potential energy = mass × change in gravitational potential

$$\Delta E_p = m\Delta V$$

The gravitational potential at the surface of the Earth is

$$V = -\frac{GM}{R} = -\frac{6.67 \times 10^{-11}\,\text{Nm}^2\,\text{kg}^{-2} \times 5.98 \times 10^{24}\,\text{kg}}{6.37 \times 10^{6}\,\text{m}}$$

$$= -6.2616 \times 10^{7}\,\text{J kg}^{-1}$$

The potential energy of a 10 kg mass on the Earth's surface is

$$10 \times -6.2616 \times 10^{7}\,\text{Jkg}^{-1} = -6.2616 \times 10^{8}\,\text{J}$$

In other words, an energy of 6.2616×10^8 J must be transferred to the mass to move it to infinity.

It is often useful to calculate the change in the potential energy.

The **gravitational potential difference** ΔV is the change in gravitational potential between two points in a gravitational field. Then the change in gravitational potential energy is given by

work done ΔW = change in gravitational potential energy = $m\Delta V$

Example

How much work would need to be done to lift a 10 kg mass to the top of Mount Everest?

Answer

We need to calculate the gravitational potential at the top of Everest, a height above sea level of about 8800 m:

$$V = -\frac{GM}{r}$$

$$= -\frac{6.67 \times 10^{-11}\,\text{Nm}^2\,\text{kg}^{-2} \times 5.98 \times 10^{24}\,\text{kg}}{(6.37 \times 10^{6}\,\text{m}) + 8800\,\text{m}}$$

$$= -6.2530 \times 10^{7}\,\text{Jkg}^{-1}$$

The difference in gravitational potential ΔV between sea level and the top of Everest is

$$(-6.2530 + 6.2616) \times 10^{7} = 8.60 \times 10^{4}\,\text{Jkg}^{-1}$$

So the energy required is $m\Delta V = 10 \times 8.60 \times 10^4 = 8.60 \times 10^5$ J.

Essential Notes

The gravitational potential has only changed a little, from sea level to the top of Everest. If we had assumed that the field was uniform, we could have used the expression $\Delta E_p = mg\,\Delta h$, which would have given the answer $10 \times 9.81 \times 8800 = 8.632 \times 10^5$ J, a difference of only 0.5%.

Gravitational field strength and gravitational potential

It is easy to get gravitational field strength and gravitational potential confused. Field strength is the *force* per unit mass, whereas potential is the *energy* per unit mass. The two quantities are connected.

In any gravitational field, the greater the field strength, the more the potential changes with distance.

Definition

The rate of change of potential with distance is equal to the field strength: $g = -\Delta V/\Delta r$.

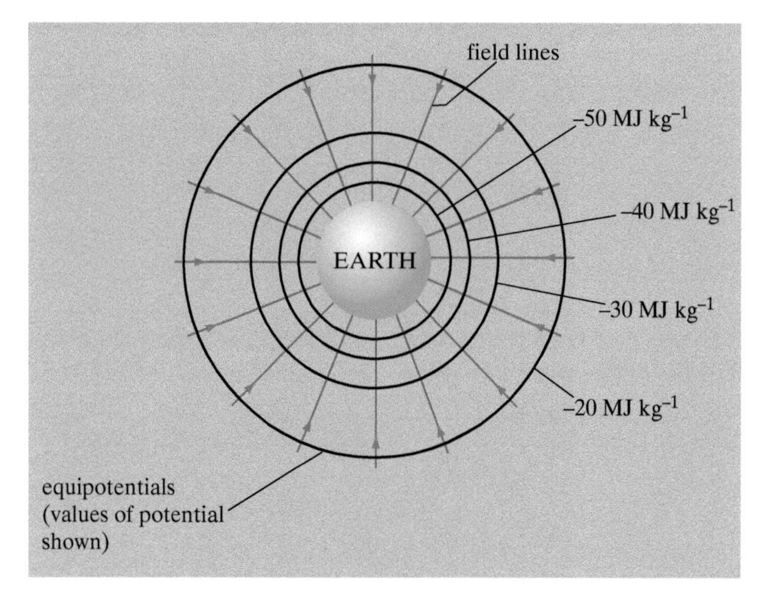

Fig 46
Field lines and equipotentials (surfaces of equal potential) around the Earth

Around a spherical mass the equipotentials are spherical. The field lines, showing the direction of the force, cross the equipotentials at right angles.

Fig 47
Contour lines connect places of the same height, and therefore the same potential.

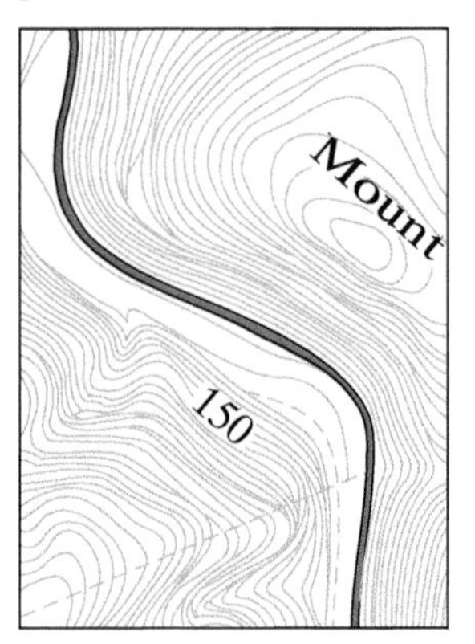

Equipotentials are similar to contour lines on a map (Fig 47); they connect points where a mass would have the same potential energy.

Close to the Earth's surface, A in Fig 48, where the gravitational potential is changing quickly with distance, the gravitational field strength is strong. Further away, at B, $\Delta V/\Delta r$ is less, and so g is smaller.

Fig 48
Potential of the Earth's gravitational field

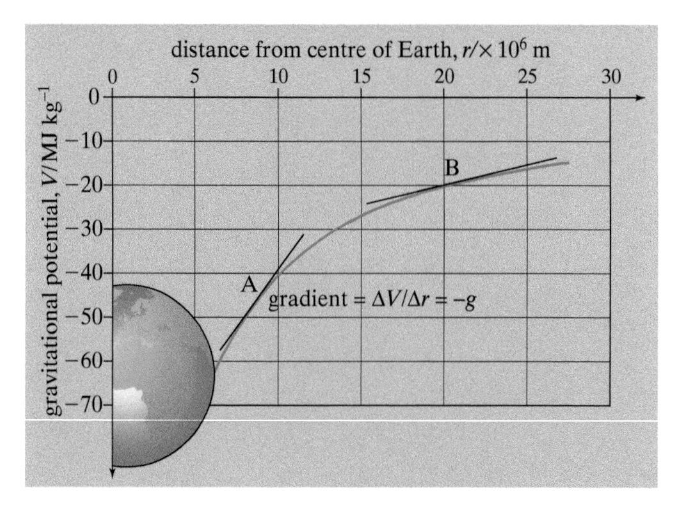

Example

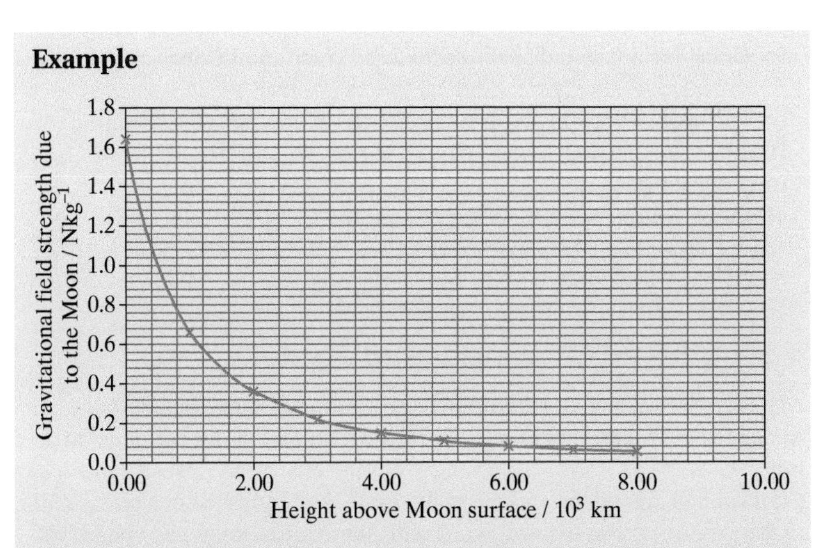

Writing about his journey back from the Moon, the Apollo 11 astronaut, Michael Collins said, "NASA (Houston) reported...the instant we left the lunar sphere of influence". He meant that they had just passed through the point where the gravitational attraction from the Earth and that from the Moon are equal and opposite, so the resultant gravitational force is zero (called the 'neutral point').

(a) Calculate the distance from the Moon to the neutral point. (Use radius of Moon = 1730 km; Earth–Moon distance between line of centres = 3.84×10^8 m; Earth is approximately 81 times more massive than the Moon.)

(b) Explain why this point was significant for the Apollo 11 astronauts.

(c) The graph shows how the gravitational field strength of the Moon varies with height above the lunar surface. How much energy would be needed to move the Apollo 11 spacecraft from the Moon's surface to the neutral point? (Mass of spacecraft = 5000 kg).

Answer

(a) Equate the gravitational field strength due to the Moon with that due to the Earth:
$Gm / r^2 = GM / (R - r)^2$, where M = mass of the Earth, m = mass of the Moon, R = Earth–Moon distance, r = distance from the centre of the Moon.
$m / M = r^2 / (R - r)^2$, so $1 / 81 = r^2 / (R - r)^2$
Hence $R^2 - 2Rr + r^2 = 81r^2$, or $80r^2 + 7.68 \times 10^8 r - 1.47 \times 10^{17} = 0$
Use the formula for solving quadratic equations to find r. This gives 38 300 km. However, this is from the centre of the Moon. Subtract the Moon's radius, 1730 km to give distance from surface of Moon ≈ 36 600 km.

(b) Once this point had been reached the gravitational attraction of the Earth took over. It was downhill all the way home!
(c) The area below the graph is the change in gravitational potential. We need the area between $x = 0$, that is from the surface of the Moon, to $x = 36\ 000$ km.
This area, estimated from the graph is $\approx 2.2 \times 10^6$ J kg^{-1}. The mass of the spacecraft was 5 000 kg, so the energy needed $\approx 5000 \times 2.2 \times 10^6 = 11$ GJ.

3.7.2.4 Orbits of planets and satellites

As a mass moves in a gravitational field its velocity changes due to the force of gravity. The exact shape of the path that it follows depends on its initial velocity, as well as the strength of the field. If you release a mass, like a cricket ball, above the surface of the Earth, it simply falls in a straight line to the ground. If you throw the ball with some initial velocity, its path will be a **parabola**.

Fig 49
Motion under the influence of gravity

The motion of a mass in the Earth's gravitational field depends on its initial velocity.

If we consider faster-moving objects, like a spacecraft approaching a planet, the general shape of the path is a hyperbola (Fig 50). If the spacecraft is moving slowly enough, it will orbit the planet. The general shape of the orbit is an ellipse. At one particular speed the orbit will be circular. When this happens, the gravitational force between the spacecraft and the planet is equal to the centripetal force.

When a satellite moves in a circular orbit, the centripetal force is provided by gravity. For a satellite of mass m moving in an orbit of radius r around a planet of mass M, this gives

$$F = \frac{GMm}{r^2} = \frac{mv^2}{r}$$

So $$v^2 = \frac{GM}{r}$$

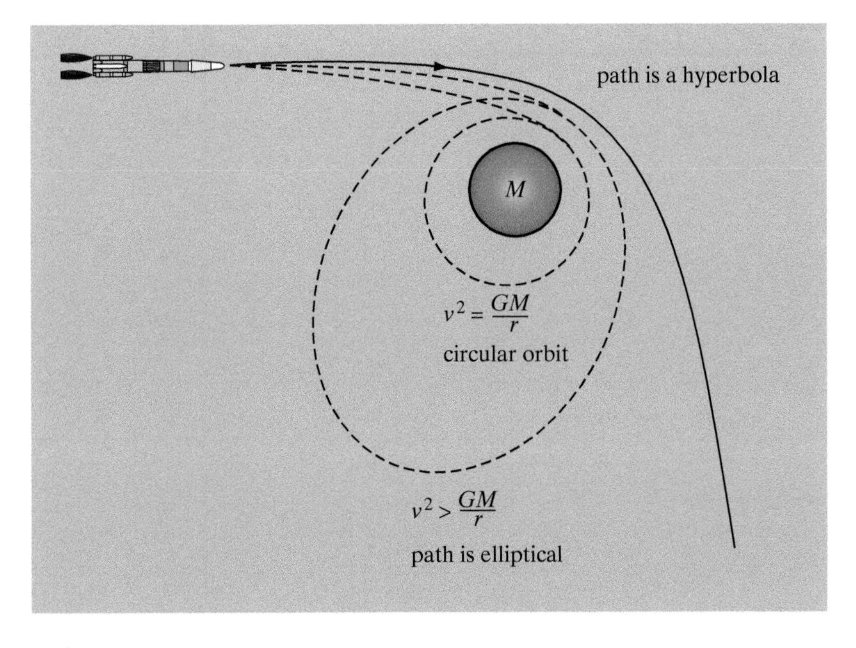

Fig 50
Motion in a gravitational field

Notes

The general path of a satellite in an orbit, such as that taken by planets around the Sun, is elliptical. However, in this specification only the special case of circular orbits are studied.

and

$$v = \sqrt{\frac{GM}{r}}$$

The orbital speed depends on the radius of the orbit and the mass of the planet, but not on the mass of the satellite.

Artificial satellites orbiting Earth

There are over 1000 operational artificial satellites orbiting the Earth. They fulfil a range of functions, from navigation (GPS) to communications. Those in a low orbit have a higher orbital speed. Satellites used for Earth observations, such as weather satellites, typically have an orbital period of 90 minutes. Satellites used for relaying certain types of telecommunications and television pictures have to be in a fixed position relative to the Earth, so that domestic satellite dishes don't have to 'track' the satellite across the sky. These **geostationary** satellites have an orbital period of 24 hours and are placed in a very high equatorial orbit. These orbits are said to be **synchronous**, meaning their orbital period matches that of the Earth's rotation.

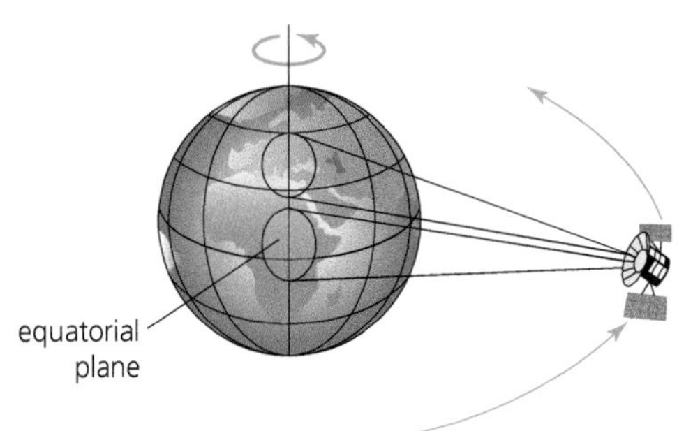

Fig 51
Geostationary satellite use for communications

Essential Notes

You need to remember the steps of this derivation, in which we show T^2 is proportional to r^3. This is true for all orbiting objects, not just geostationary artificial satellites.

We can determine the radius of a geostationary orbit using the following derivation. Take the Earth's mass as 5.98×10^{24} kg and G as 6.67×10^{-11} N m^2 kg^{-2}.

Assuming that the orbit is circular,

$$F = \frac{GMm}{r^2} = \frac{mv^2}{r}$$

which gives

$$\frac{GM}{v^2} = r \qquad (1)$$

The time taken for one orbit is

$$T = \frac{\text{distance}}{\text{speed}} = \frac{2 \times \pi \times r}{v}$$

which gives

$$v = \frac{2\pi r}{T} \qquad (2)$$

Combining equations (1) and (2) to eliminate v,

$$\frac{GM}{(2\pi r/T)^2} = r \quad \text{and so} \quad \frac{GMT^2}{4\pi^2} = r^3$$

For a satellite orbiting Earth in a geostationary orbit, the orbital period has to be 24 hours: $T = 24 \times 60 \times 60 = 86\,400$ s

$$\text{So } r^3 = \frac{6.67 \times 10^{-11} \times 5.98 \times 10^{24} \times 7.465 \times 10^9}{39.48} = 7.54 \times 10^{22}\,\text{m}^3$$

This gives a value for r of 42.2×10^6 m.

Since the Earth's radius is 6.38×10^6 m, a geostationary satellite has to be placed in an orbit that is approximately 6 Earth radii above the Earth's surface.

Example

Johannes Kepler used the detailed observations of Tycho Brahe to propose his laws of planetary motion. Kepler deduced that the orbits of the planets were ellipses and that the period of the orbit, T, was connected to the mean radius, r, by the relationship $T^2 \propto r^3$ (Kepler's third law).

(a) Assume that the orbits of the planets are circular and use Newton's law of gravitation to derive Kepler's third law.

(b) Explain how you could use the data in Table 4 to plot a graph, with logarithmic axes, to verify Kepler's third law.

(c) Plot the graph and explain whether it does verify Kepler's third law.

Table 4

	Mercury	Venus	Earth	Mars	Jupiter	Saturn	Uranus	Neptune
Mean distance from Sun/AU	0.39	0.72	1	1.52	5.20	9.54	19.18	30.06
Orbital period/ Earth years	0.24	0.62	1	1.88	11.86	29.46	84.01	164.8

Essential Note

AU is an Astronomical Unit, the mean Earth–Sun distance.

Answers

(a) Gravitation provides the centripetal force so that $GMm / r^2 = mr\omega^2$, where M = solar mass, m = planetary mass, r = mean radius, ω = angular frequency of the orbit = $2\pi f = 2\pi / T$.

So $GMm / r^2 = mr4\pi^2 / T^2$, simplifying this gives $GMmT^2 = mr4\pi^2r^2$, or $GMT^2 = 4\pi^2r^3$.

So $T^2 = (4\pi^2 / GM)\, r^3$. Since $4\pi^2 / GM$ is a constant, $T^2 \propto r^3$.

(b) Since $T^2 = (4\pi^2 / GM)\, r^3$ a graph of T^2 (y-axis) against r^3 (x-axis) should be a straight line. Or we could plot a graph of log T against log r.

$\log T = (3/2) \log r + ½ \log (4\pi^2 / GM)$. A graph of log T (y-axis) against log r (x-axis) should have a gradient of 3/2.

(c)

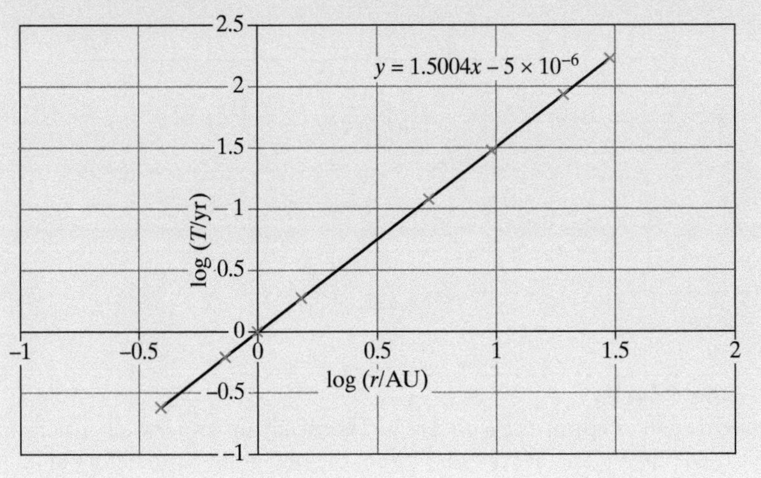

Essential Note

Notice how the label on a logarithmic axis is written, e.g. log (T/yr). A log has no units but the original variable usually does.

The graph is a straight line with a gradient of 1.5, therefore, $T \propto r^{3/2}$.

The energy of an orbiting satellite

The total energy of a satellite is the sum of its potential energy and its kinetic energy. If the satellite is moving at a velocity v in a circular orbit of radius r around the Earth then:

$$E_{total} = E_k + E_p = \tfrac{1}{2}mv^2 - \frac{GMm}{r}$$

For a circular orbit, equating the gravitational force to the centripetal force,

$$\frac{mv^2}{r} = \frac{GMm}{r^2}$$

So, substituting,

$$E_{total} = \frac{GMm}{2r} - \frac{GMm}{r} = -\frac{GMm}{2r}$$

We can apply this to a satellite that is in a decaying orbit, spiralling towards the Earth. As it gets closer to Earth, r decreases and E_{total} becomes more negative (the total energy decreases). However, the decrease in total energy is only half the decrease in potential energy. The difference is due to the increase in kinetic energy. The satellite speeds up as it loses potential energy. If nothing is done to stop it, it will crash at high speed into the Earth.

Fig 52
Orbital energy

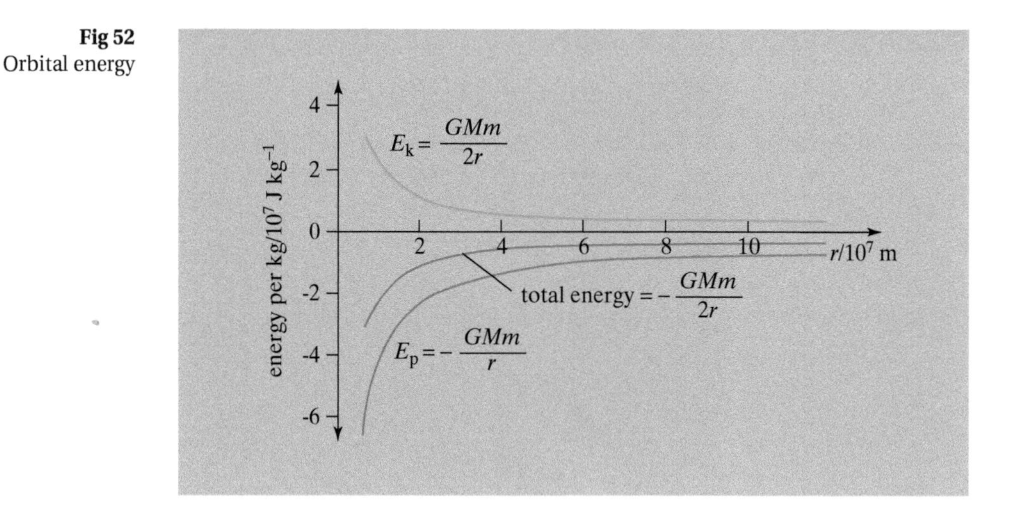

Escape velocity

Imagine you are throwing a cricket ball vertically up into the air. The faster it is travelling when it leaves your hand, the higher it will go before falling back to Earth. It is possible, in theory at least, to throw the ball at such a high speed that it would never come down again. This speed is known as the **escape velocity**. We can calculate it from the value of the gravitational potential at the Earth's surface, E_{PEarth}

$-E_{PEarth}$ = energy needed to move a mass to infinity.

If this all came from the initial kinetic energy of the mass then, ignoring any work done against drag,

$$\Delta E_k = -\Delta E_p$$

For a mass on the Earth:

$$\tfrac{1}{2}mv^2 = \frac{GMm}{R}$$

So:

$$\tfrac{1}{2}v^2 = \frac{GM}{R}$$

The escape velocity is therefore:

$$v = \sqrt{\frac{2GM}{R}}$$

For an object on the surface of the Earth where $g = 9.81\ \mathrm{N\,kg^{-1}}$, $v = 11\ \mathrm{km\,s^{-1}}$. If you could throw an object, of any mass, up at this speed, it would never come down.

Essential Notes

You don't have to reach escape velocity to leave the Earth. You could walk to the Moon if you had a long enough ladder. The escape velocity is the velocity required when all the energy has to come from the initial kinetic energy. Space rockets don't need to travel this fast, because they transfer energy in several bursts as they move further away from the Earth.

3.7.3 Electric fields

3.7.3.1 Coulomb's law

There is an electrostatic force that acts between all electric charges. This force is described by **Coulomb's law**. Coulomb's law states that the force, F, between two charges, Q_1 and Q_2, which are separated by a distance r is:

- proportional to the product of the two charges, $F \propto Q_1Q_2$;
- inversely proportional to the square of the distance between the charges, $F \propto 1/r^2$.

This force law is similar to Newton's law of gravitation (page 50). Both laws follow the inverse-square relationship. One difference is that the force between two charges depends on the medium between the charges, whereas the gravitational force is independent of the medium. For two charges in a vacuum the force between them is

$$F = \frac{Q_1Q_2}{4\pi\varepsilon_0 r^2}$$

The constant ε_0 is known as the **permittivity of free space**. The value of ε_0 is $8.85 \times 10^{-12}\ \mathrm{F\,m^{-1}}$. The force between two 1 coulomb charges placed 1 m apart in a vacuum is therefore

$$F = \frac{1}{4\pi\varepsilon_0} = 9 \times 10^9\ \mathrm{N}$$

The electrostatic force can be attractive or repulsive. Two charges of similar sign, two positives or two negatives, will repel each other. The force between two opposite charges, positive and negative, will be attractive. An attractive force is given a negative sign.

For an electrically charged sphere, we can consider the charge to be concentrated at the centre of the sphere.

Essential Notes

If the two charges are separated by another medium, such as air, the force between them is reduced and the equation becomes

$$F = \frac{Q_1Q_2}{4\pi\varepsilon r^2}$$

ε is the permittivity of the material, which is usually given in terms of how much bigger it is than the permittivity of free space, $\varepsilon = \varepsilon_r\varepsilon_0$, where ε_r is the relative permittivity.

ε_r for air is 1.005, so we can usually neglect the effect of air.

Example

A hydrogen atom consists of a single electron orbiting a single proton. Compare the size of the gravitational and electrostatic forces between the particles.

Data:	Mass of proton	1.67×10^{-27} kg
	Mass of electron	9.11×10^{-31} kg
	Charge on an electron	-1.60×10^{-19} C
	Charge on a proton	1.60×10^{-19} C
	Radius of a hydrogen atom	5.29×10^{-11} m

Answer

The size of the electrostatic force is

$F = Q_1Q_2/4\pi\varepsilon_0 r^2 = 8.23 \times 10^{-8}$ N

The size of the gravitational force is

$F = G\,m_1m_2/r^2 = 3.63 \times 10^{-47}$ N

The electrostatic force between the proton and the electron is about 2×10^{39} times larger than the gravitational force.

3.7.3.2 Electric field strength

Every charged particle creates an electric field around itself. An electric field is a region of space where a charged particle experiences a force. We can visualise electric fields by sketching the field lines around a charge. The field lines show the direction of the force that would be exerted on a positive charge placed in the field.

Around a point charge, or a spherical charge distribution, the field lines are radial. See Fig 53, opposite.

The strength of the electric field, E, is defined as the force, F, that would be exerted on a unit charge. Field strength is measured in newtons per coulomb, N C^{-1}.

Definition

Electric field strength, E, *is the force per unit charge:* E = F/Q.

Essential Notes

Electric field strength is a vector quantity. Its direction is that of the force acting on a positive charge.

Since the force between two point charges, Q and q, is $F = Q\,q/4\pi\varepsilon_0 r^2$, the electric field strength in a radial field is

$$E = \frac{F}{q} = \frac{Q}{4\pi\varepsilon_0 r^2}$$

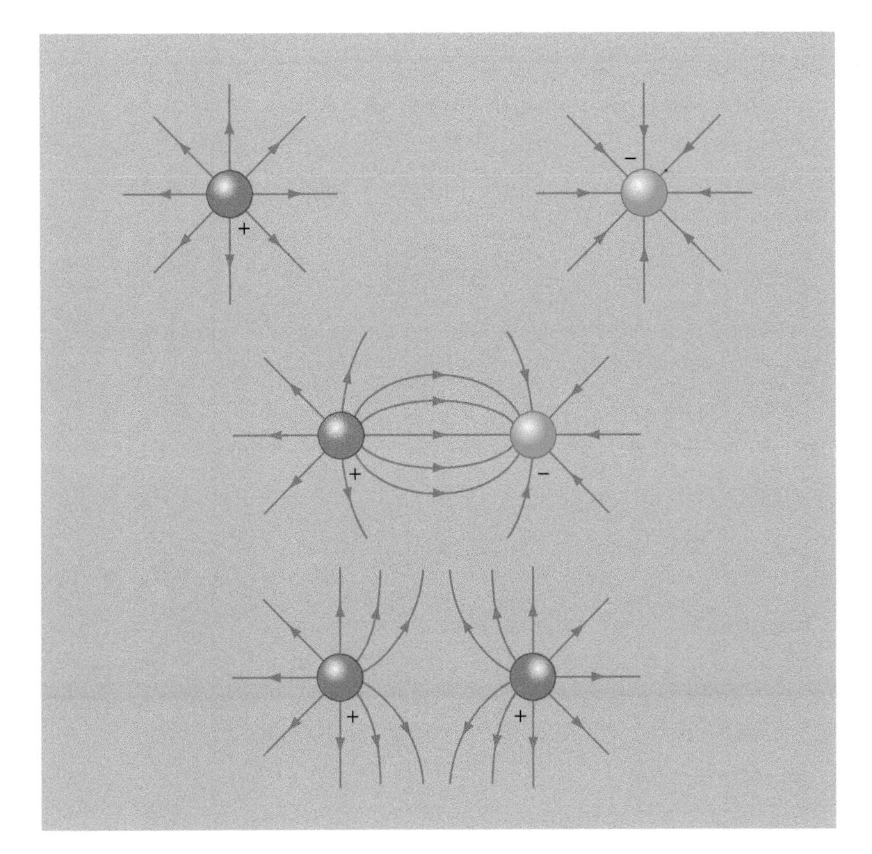

Fig 53
Field lines around charges

The field lines show the direction of the force on a positive charge; away from another positive charge, and towards a negative charge.

Uniform field

It is possible to create a uniform electric field between two oppositely charged parallel conductors, such as the plates of a capacitor. In between the plates the field strength is independent of position and just depends on the separation of the plates, d, and the **potential difference**, V, between them.

Essential Notes

Remember that potential difference, V, between two points is defined as the work done (or energy transferred), ΔW, per unit charge moved between the points:

$$\Delta V = \Delta W / Q$$

and is measured in joules per coulomb, J C^{-1}, or volts.

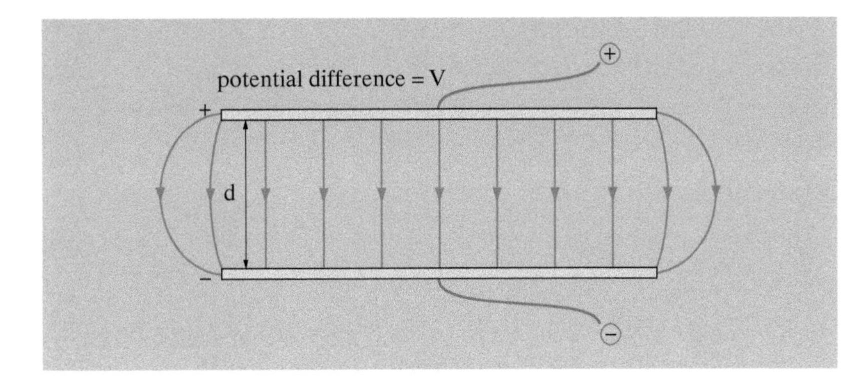

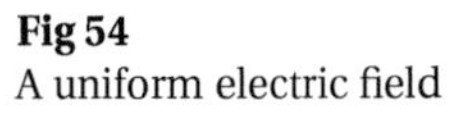

Fig 54
A uniform electric field

The electric field strength in a uniform field is

$$E = \frac{V}{d}$$

Notes

Electric field strength can be measured in units of N C^{-1} or V m^{-1}.

Essential Notes

There is no electric field inside a closed conductor. A hollow metal container, known as a Faraday cage, may have charges on the outside but will not have any charges on its inside surface. Faraday cages are used to shield people or sensitive equipment from intense electric fields.

We can show that the magnitude of a uniform electric field, E, between two parallel plates depends on the potential difference, ΔV, between the plates and their separation, d, in the following way.

The work done, W, in moving a charge, Q, through a potential difference of ΔV is given by $W = Q\,\Delta V$. But, for a constant force F, the work done is force × distance, Δd, moved:

$$W = F\Delta d$$

Since this is a uniform field, the force on the charge does not depend on the charge's position between the plates; i.e., F is constant.

Therefore, $Q\,\Delta V = Fd$, or $F = Q\,\Delta V / d$.

The electric field, E, is defined as the force per unit charge, $E = F / Q$.

Therefore:

$$E = \Delta V / d$$

Example

An electron is placed between two metal plates, which are 10 cm apart in a vacuum. A potential difference of 100 V is applied across the plates. Calculate the magnitude of the force on the electron.

Answer

The electric field strength in this uniform field is

$$E = \frac{V}{d} = \frac{100}{0.1} = 1000\,\text{V}\,\text{m}^{-1}$$

Field strength is the force on a unit charge, so the force on a charge q is $F = qE$.

The charge on an electron is -1.6×10^{-19} C. So the magnitude of the force on the electron is

$F = 1.6 \times 10^{-19} \times 1000 = 1.6 \times 10^{-16}$ N

3.7.3.3 Electric potential

Electric potential is defined in a similar way to gravitational potential (page 51).

Definition

*The **electric potential**, V, at a point in an electric field is defined as the work done in bringing a unit positive charge from infinity to that point.*

The potential at a point in an electric field is the potential energy of a unit charge placed at that point. Electric potential is measured in joules per coulomb, J C^{-1}, or volts.

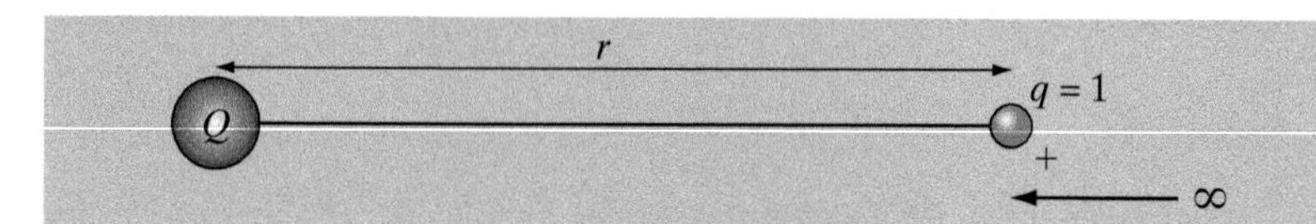

Fig 55
The definition of potential

The force on a charge in a radial field is not constant but obeys Coulomb's law. The work done in moving a unit positive charge ($q = 1$) through a small distance Δr is

$$\Delta W = F\,\Delta r = Q\,\Delta r / 4\pi\varepsilon_0 r^2$$

This is the area of the small shaded strip in Fig 56. The area below the force curve is equal to the total work done in bringing a unit positive charge from infinity to that point, so electric potential in a uniform field at a distance r from a charge Q is given by

$$V = \frac{Q}{4\pi\varepsilon_0 r}$$

Essential Notes

Electric potential is a scalar quantity; it has no associated direction.

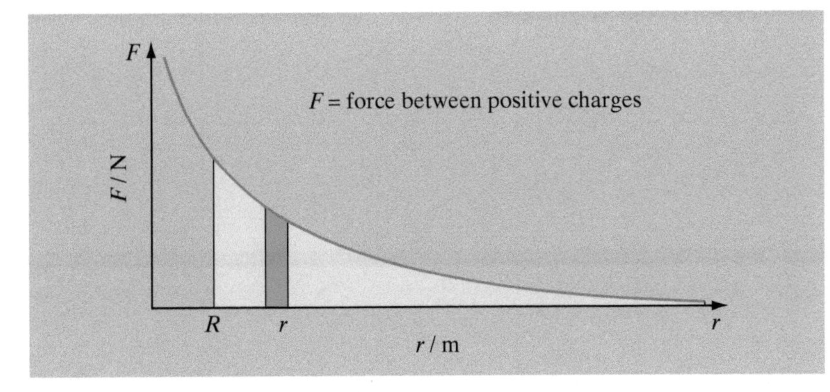

Fig 56
Work done in an electric field

The total work done in bringing a charge from infinity to R is the light blue area under the curve.

The **potential difference** between two points in an electric field is the work done (energy transferred) for a unit charge moved between the point and so is measured in J C^{-1}, or volts. The actual work done depends on the size of the charge:

$\Delta W = Q\,\Delta V$

If two positive charges are moved closer together, positive external work has to be done to overcome the repulsive force. So the potential around a positive charge is defined as being positive. The negative potential around a negative charge indicates that work would need to be done to pull a positive charge **away**, towards infinity. (This is similar to the potential in a gravitational field where the force is attractive and the potential is always negative.)

Essential Notes

Suppose we move a charge of just one electron, rather than a whole coulomb of charge, through a potential difference of 1 volt. The work done then is

$W = qV = 1.6 \times 10^{-19}$ J

This small amount of energy is known as an **electron volt, eV.**

Notes

Potential is inversely proportional to r, whereas field strength is inversely proportional to the **square** of r.

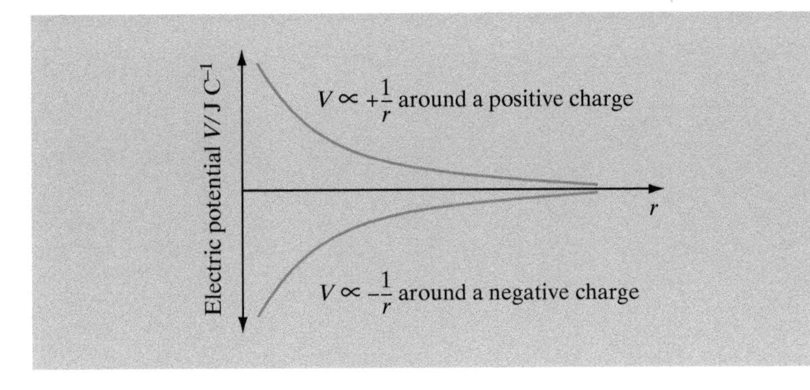

Fig 57
Electric potential in the field around a positive and negative charge

Essential Notes

Potential drops to zero at infinity, but infinity is not a very practical reference point. We often use the Earth's surface as zero potential. The Earth is a reasonably good conductor and is therefore all at the same potential.

Example

In Rutherford scattering an alpha particle approaches a gold nucleus. Calculate the work done by the alpha particle as it moves from a large distance away to within 0.5×10^{-12} m from the gold nucleus. (The proton number of gold is 79.)

Answer

The electric potential at 0.5×10^{-12} m from the gold nucleus is

$$V = \frac{Q}{4\pi\varepsilon_0 r}$$

$$= \frac{79 \times 1.6 \times 10^{-19}}{4\pi \times 8.85 \times 10^{-12} \times 0.5 \times 10^{-12}}$$

$$= 2.27 \times 10^{5}\ \mathrm{J\,C^{-1}}$$

Since the charge on the alpha particle is $2 \times 1.6 \times 10^{-19}$ C, the work done is

$$W = qV = 2 \times 1.6 \times 10^{-19} \times 2.27 \times 10^{5} = 7.27 \times 10^{-14}\ \mathrm{J}$$

(Supposing this energy comes from the kinetic energy of the alpha particle, the particle would have to be emitted with an initial energy of 7.27×10^{-14} J = 0.45 MeV.)

Equipotential surfaces

The work done, W, in moving a charge, Q, through a potential difference, ΔV, is given by $W = Q\,\Delta V$. But it is possible for a charge to follow a trajectory where the potential does not change, so $\Delta V = 0$.

A surface that links points of equal potential is known as an **equipotential surface,** and no work is done by a charge moving on such a surface. In a conductor, a metal sheet for example, charge can flow until the potential is the same throughout the conductor. A conductor is an equipotential surface.

In fact the field strength, E, is equal to the potential gradient, $\Delta V/\Delta r$:

$$E = \Delta V/\Delta r$$

In a strong electric field, the equipotential surfaces are close together.

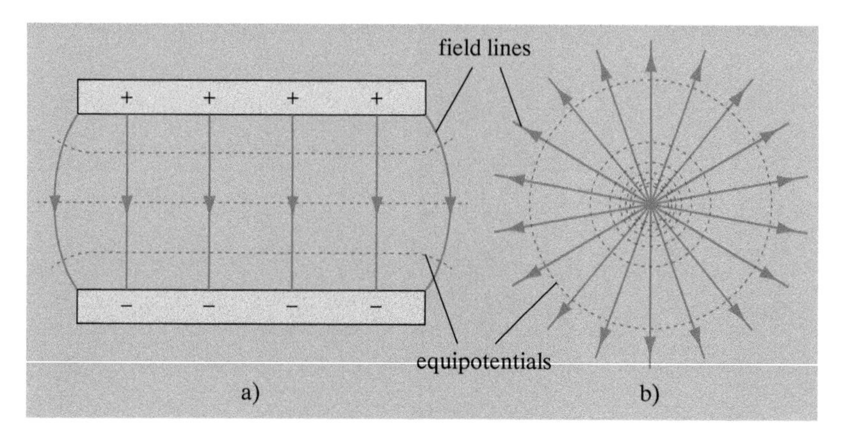

Field lines and equipotentials for (a) a uniform field and (b) a radial field; note that the field lines always cross the equipotential lines at right-angles.

Essential Notes

Equipotentials in electric fields are similar to those in gravitational fields (page 54). They connect points where a charge has the same potential energy.

Motion of charged particles in a uniform electric field

In a uniform electric field, charged particles accelerate in a similar way to masses moving in a gravitational field. In a uniform field the force, F, on a charge, q, is constant: $F = qE$. The direction of the force on a charge is also constant, with the result that the particle will move in a parabolic path.

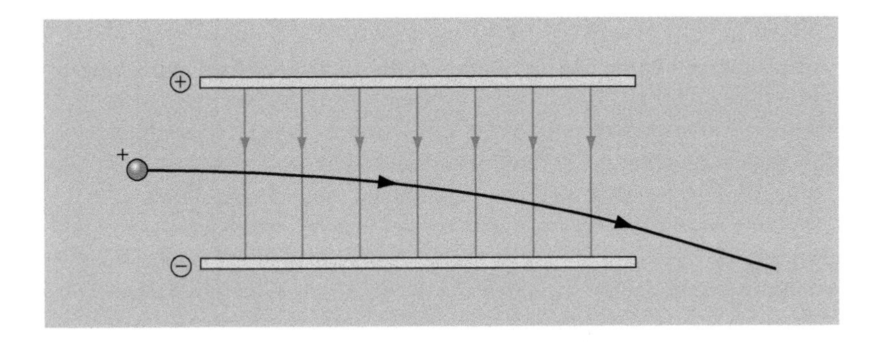

Fig 58
Charged particle moving in a field A positive charge is attracted towards the negative electrode. The particle moves in a straight line when outside the electric field between the charged plates; its path follows the curve of a parabola within the electric fields.

Comparison of electric and gravitational fields

Gravitational and electric fields both have force laws which obey the inverse-square law. The major differences arise because there are positive and negative charges, but only positive masses. The gravitational force is always attractive, whereas the electric force may be attractive or repulsive. The gravitational force is independent of the medium between the masses, whereas the force between two charges is affected by the material between them.

Table 5
Gravitational force and electrostatic force compared

	Gravitational force	Electric force
Force law	Newton's law of gravitation $F = G\,m_1m_2/r^2$	Coulomb's law $F = Q_1Q_2/4\pi\varepsilon r^2$
Force proportional to	Product of masses $1/r^2$	Product of charges $1/r^2$
Constant of proportionality	G This is a universal constant. The force is independent of the medium	$1/4\pi\varepsilon$ The force depends on the permittivity of the intervening medium
Type of force	Always attractive	Can be attractive or repulsive
Field strength (radial)	$g = GM/r^2$ Force on unit mass, N kg^{-1}	$E = Q/4\pi\varepsilon r^2$ Force on unit charge, N C^{-1}
Potential (radial field)	$V = -GM/r$ Potential energy per unit mass, J kg^{-1}	$V = Q/4\pi\varepsilon_0 r$ Potential energy per unit charge, J C^{-1}

3.7.4 Capacitance

3.7.4.1 Capacitance

Notes

The current in a simple capacitor circuit, as in Fig 59, changes from a maximum value when the capacitor is uncharged, to zero when the potential difference across the capacitor is equal to the e.m.f. of the battery.

Capacitors are devices that store electric charge. At their simplest they consist of two metal plates separated by an air gap. When a capacitor is connected to a battery, negative charges (electrons) flow on to one plate and off the other plate. As the charge on the plates increases, it becomes more difficult for electrons to flow onto the negative plate, or to flow off the positive plate (see Fig 59), and the current in the circuit drops. Eventually the potential difference across the capacitor is equal to the potential difference across the battery and no more current flows. The capacitor is now fully charged, with a charge of $+Q$ on one plate and a charge of $-Q$ on the other plate. It would be possible to store more charge on the capacitor by increasing the potential difference across the capacitor.

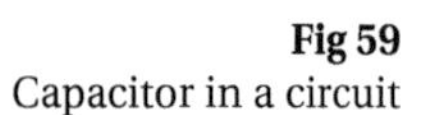

Fig 59
Capacitor in a circuit

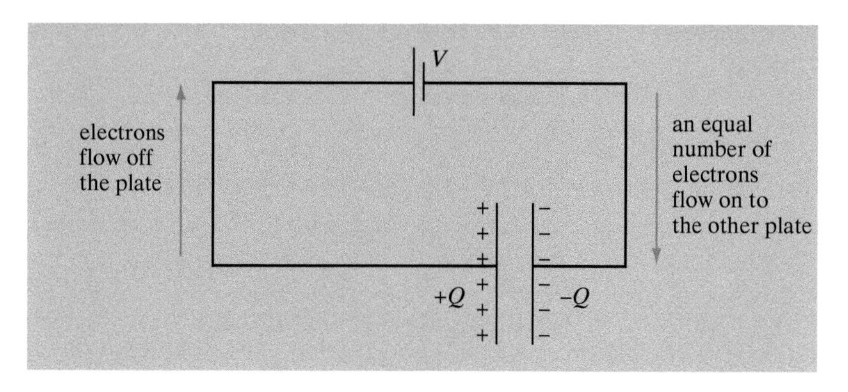

The charge stored, Q, is proportional to the potential difference, V, across the capacitor:

$$Q \propto V$$

We can write this as

$$Q = CV$$

where C is a constant known as the **capacitance**.

Notes

The unit of capacitance, the farad, is very large. Real capacitors tend to have values measured in millifarads, mF, microfarads, μF, nanofarads, nF, or picofarads, pF.

The capacitance is a constant for a given capacitor and its value depends on the construction of the capacitor, in particular the area and separation of the plates, and the material used between the plates.

If we rearrange the equation $Q = CV$, we can define capacitance:

$$C = Q/V$$

as the charge stored on the capacitor, per unit potential difference across the capacitor. In SI units, capacitance is measured in coulombs per volt. $1\ \mathrm{C\,V^{-1}}$ is known as a **farad, F**.

Definition

Capacitance is the charge stored by a capacitor, per volt of potential difference applied across it. A capacitance of one farad will store one coulomb of charge for every volt applied across the capacitor.

We could store more charge on the capacitor by increasing the potential difference across it. However, there is a limit to this. Eventually the potential difference across the insulator (between the capacitor plates) will be so great that electrical breakdown will occur and conduction will take place across the gap. A capacitor is often labelled with its capacitance in farads, and its maximum working potential difference in volts.

Example

A capacitor is labelled 100 nF 10 V. Calculate the maximum charge that can be stored on the capacitor.

Answer

The maximum potential difference that can be applied across the capacitor is 10 V. The maximum charge stored will then be

$Q = CV = 100 \times 10^{-9}\ \text{F} \times 10\ \text{V} = 1 \times 10^{-6}\ \text{C}$

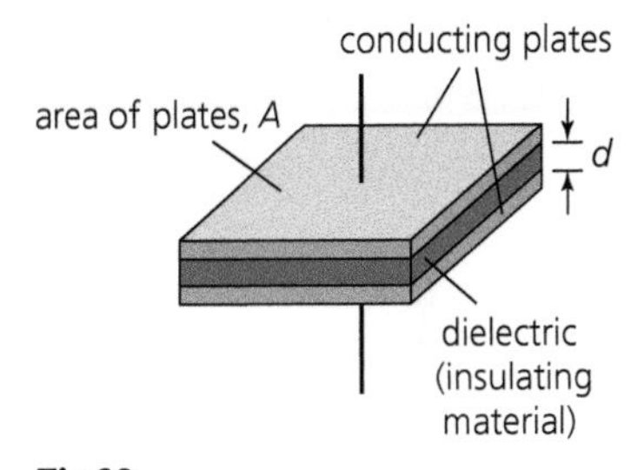

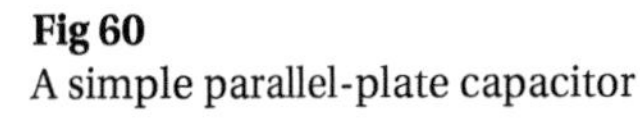
Fig 60
A simple parallel-plate capacitor

3.7.4.2 Parallel plate capacitor

A simple form of capacitor is two conducting plates, each of area A, separated by a non-conducting material, of thickness d (Fig 60). When a potential difference of V is placed across the plates, the amount of charge that can be stored on the plates, Q, depends on the physical characteristics of the capacitor. The capacitance, C, of such a capacitor depends on:

- Area, A – the larger the area of the plates, the more charge can be stored, so the greater the capacitance.
- Separation, d – The closer the plates are to each other, the greater the capacitance.
- Permittivity, ε – The **dielectric** material between the plates is an insulator; the charges are not free to move. However, the molecules in a dielectric material can align themselves with the electric field between the plates (Fig 61). This reduces the force that opposes charge being added to the capacitor.

If there were a vacuum between the plates, then the permittivity would be that of free space, $\varepsilon_0 = 8.85 \times 10^{-12}\ \text{F m}^{-1}$. Materials that have molecules which are highly polar have a higher value of permittivity; one example of such a material is water. The **relative permittivity**, ε_r, is the permittivity as compared to that of free space. For example, the relative permittivity of water is 80, which means that the permittivity of water, $\varepsilon = 80 \times 8.85 \times 10^{-12} = 7.1 \times 10^{-10}\ \text{F m}^{-1}$. In general, $\varepsilon = \varepsilon_r\, \varepsilon_0$.

So the capacitance of a parallel plate capacitor is given by $C = \varepsilon A / r$.

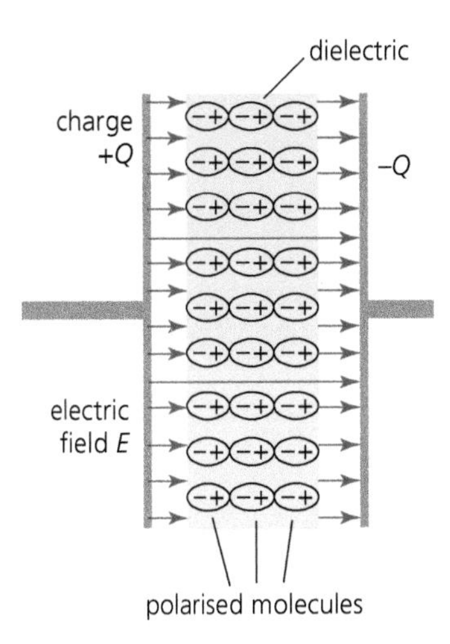

Fig 61
The **polar molecules** in the dielectric align themselves with the applied electric field allowing more charge to flow onto the capacitor.

Example

A parallel plate capacitor is made from two large metal discs of radius 20 cm. They are placed 1 mm apart in air. $\varepsilon_r = 1.0005$, breakdown voltage of air is approximately equal to 3×10^6 V m^{-1}.

(a) Calculate the capacitance of the arrangement.

(b) Estimate the maximum charge that could be stored on this capacitor.

(c) Suggest two ways of increasing the charge stored.

Answer

(a) $A = \pi r^2 = \pi \times 0.20^2 = 0.126$ m^2.

$C = 0.126 \times 1.0006 \times 8.85 \times 10^{-12} / 0.001 = 1.1$ nF.

(b) The largest potential difference that can be applied across a 1 mm gap in air = 3 kV. $Q = CV = 1.1 \times 10^{-9} \times 3000 = 3.3$ μC.

(c) Use a material with a higher relative permittivity between the plates, water or plastic for example. Increase the area of plates.

Decreasing the separation of the plates or increasing the potential across the plates, would lead to breakdown, unless a material with a higher breakdown potential is placed between the plates.

3.7.4.3 Energy stored by a capacitor

A capacitor stores electric charge. It is also, therefore, a store of electrical energy. The work done in charging a capacitor is stored until the capacitor is discharged, when the electrical energy may be usefully transferred, to drive a motor or light a bulb for example. The amount of energy that is stored can be calculated by considering the charging process.

A battery can be used to charge a capacitor (Fig 59). As a charge Q flows onto the capacitor, an amount of work ΔW has to be done to overcome the potential difference V, due to the charge that is already on the capacitor. The work done ΔW in moving a charge Q through a potential difference ΔV is

$$\Delta W = Q\Delta V$$

(See page 63 for the definition of the volt.)

But as each charge flows onto the capacitor, the potential difference increases. The work done in adding the next charge will be greater (see Fig 62). If the capacitor is initially uncharged, and is then charged until the potential difference is equal to V, the mean (average) potential difference is given by

mean potential difference = (final p.d. + initial p.d.) / 2

$$= \frac{V + 0}{2} = \tfrac{1}{2}V$$

So the total work done in charging the capacitor is:

$$W = Q \times \tfrac{1}{2}V = \tfrac{1}{2}QV$$

This is also equal to the energy stored by the capacitor, E. Since $Q = CV$, we can write:

$$E = \tfrac{1}{2}QV = \tfrac{1}{2}CV^2 = \tfrac{1}{2}\frac{Q^2}{C}$$

The energy stored can also be derived from a graph of potential difference against charge for the capacitor while charging. See Fig 62.

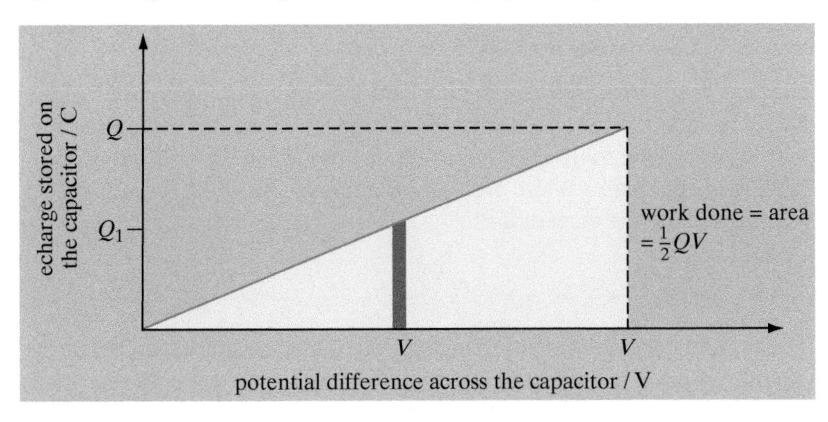

Fig 62
The charge stored by a capacitor is proportional to the potential difference across it, $Q \alpha V$, or $Q = CV$. The gradient of this graph is the capacitance, C.

The area of the shaded strip is $q\Delta V$. This is the work done in moving a charge q through a potential difference ΔV. The total work done in charging the capacitor to a charge Q and a potential difference V is the sum of the area of such shaded strips, which gives $W = \frac{1}{2}QV$.

Example

The capacitor used to store charge in a camera flash unit has a capacitance of 470 mF and can be charged to a potential of 30.0 V.

(a) How much energy is stored by the capacitor when it is fully charged?

(b) If the capacitor discharges through the flash in a time of 0.20 ms, calculate the average power.

Answer

(a) The energy stored is $E = \frac{1}{2}QV = \frac{1}{2}CV^2 = 0.5 \times 470 \times 10^{-3} \times 30^2$
$= 212$ J (to 3 s.f.).

(b) Power is rate of energy transfer $= 211.5/0.20 \times 10^{-3} = 1.06$ MW.

Notes

Charging a capacitor is rather like stretching a spring: easy at first, and then progressively more difficult as the opposing force increases. A similar formula is used to calculate the energy stored in each case:

- the work done in stretching a spring is $\Delta W = \frac{1}{2}k\,(\Delta x)^2$
- the work done in charging a capacitor is $\Delta W = \frac{1}{2}\,C\,(\Delta V)^2$

3.7.4.4 Capacitor charge and discharge

Charging a capacitor through a resistor

A capacitor can be charged by connecting it to a cell or other d.c. electrical power source. The power source has e.m.f. E. The charging circuit will have a resistance, which we can represent by a resistor R (see Fig 63). If the capacitor is initially uncharged, when the switch is closed the potential difference across the resistor will be E and the initial charging current will be $I_0 = E/R$

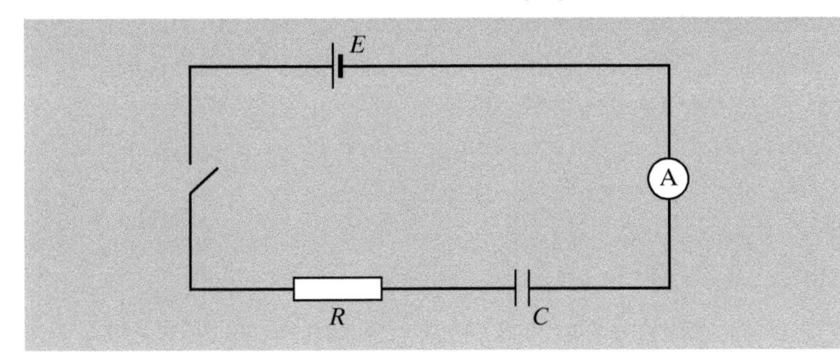

Fig 63
A charging circuit for a capacitor

The resistance R represents the total resistance of the circuit, including the internal resistance of the battery and the resistance of any connections.

However, as time progresses and charge accumulates on the capacitor, there will be a potential difference V across the capacitor. This opposes the e.m.f. of the cell, and so the potential difference across the resistance is now $E - V$. This is smaller than the initial value and so the charging current drops. Eventually the potential difference across the capacitor is equal to the e.m.f. of the cell, so that the potential difference across the resistor falls to zero and no more charge flows.

Suppose we decreased the resistance of the circuit. The charging current would be larger. The capacitor would charge up in less time, and the current would fall more quickly. In fact, the larger the current, the quicker the current decreases. More precisely, the rate of change of the current is proportional to the current itself.

$$\frac{\Delta I}{\Delta t} \propto I$$

This type of behaviour, where a change in a quantity is proportional to the quantity itself, leads to an **exponential** relationship. In this case, since the current is decreasing, the relationship is known as exponential decay.

Fig 64
Charging current against time

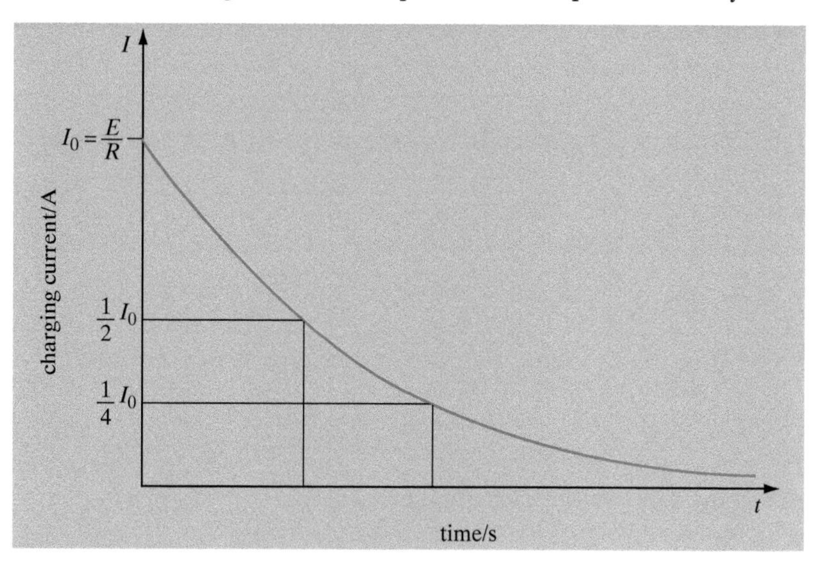

Essential Notes

The equation for an exponential decay curve has the form $y = Ae^{-bx}$, where A and b are constants. You can find the value of the number e by entering e^1 on your calculator. You should get a value of 2.718281828 (to 10 s.f.).

Notes

Since C has units of coulombs per volt (C = Q/V) and R has units of volts per amp (R = V/I), the **time constant** RC has units of:

(coulombs / volts) × (volts per ampere) = (coulombs / amperes)

But an ampere is a coulomb per second, so RC has units of seconds.

Notes

Exponential decay curves have the property that they always take the same time to decrease by a given fraction.

The charging current always follows an exponential decay curve, but the exact values followed by the curve depend on:

- The capacitance, C, of the capacitor. A larger capacitance will hold more charge and it will take longer to charge.
- The resistance, R, of the circuit. A larger resistance will reduce the charging current, and extend the time taken to charge the capacitor.

The product RC is known as the time constant. It has units of seconds.

The equation that describes the graph in Fig 64 is

$$I = I_0 \, e^{-t}/RC$$

When $t = 0$, $e^{-t}/RC = 1$, so $I = I_0$, the initial charging current.

When $t = RC$, $e^{-t}/RC = e^{-1}$. The quantity e – 1 is equal to 0.37 (to 2 s.f.), so $I = 0.37\, I_0$.

After one time constant, the current has dropped to 0.37 of its original value. After two time constants, the current will drop to e^{-2}, which is $(0.37)^2 = 0.14$ of its original value.

Definition

The time constant is the time that it takes for the current to drop to 1/e (≈ 0.37) of its original value.

Example

A 100 μF capacitor is charged from a 10 V supply in a circuit of resistance 1 kΩ.

(a) What will the initial current be?

(b) What is the time constant of the circuit?

(c) What would the current be after a time equal to four time constants has passed?

Answer

(a) $I_0 = E/R = 10/1000 = 0.01$ A = 10 mA.

(b) Time constant = $RC = 1000 \times 100 \times 10^{-6} = 0.1$ s.

(c) After four time constants, $t = 0.4$ s, the current will have dropped to $1/e^4$ of its original value:

$1/e^4 = 0.0183$

So the current will be 0.0183×10 mA = 0.183 mA.

Notes

We can make sure that a capacitor is not charged at the start of an experiment by touching its terminals together. Initial values of variables such as current and potential difference are denoted by a subscript, e.g. I_0, V_0 to indicate that this is the value at the start, time t = 0 s.

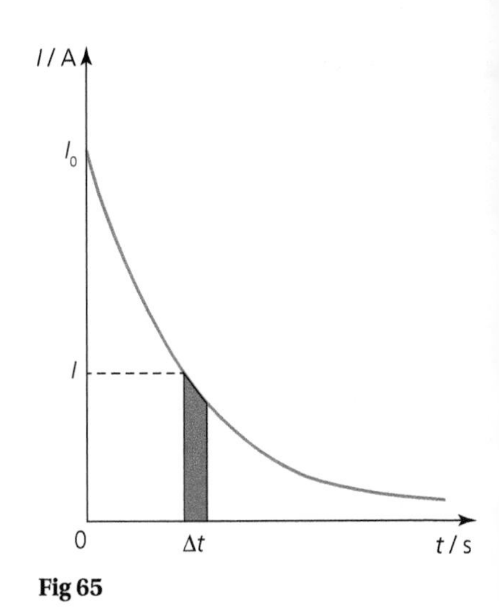

Fig 65

Charging a capacitor: potential difference and charge

As charge flows in the circuit, the capacitor becomes charged. Since current is the rate of flow of charge, the charge ΔQ that flows on to the capacitor in a short time interval Δt is equal to $I \times \Delta t$. This is equivalent to the area of the red shaded strip in Fig 65. The total area below the graph (the sum of the area of all the strips) is the total charge on the capacitor. This will eventually be equal to the steady state value, $Q = CV$. A graph showing how charge on the capacitor changes with time is shown in Fig 66.

The potential difference across the capacitor is proportional to the charge stored, $Q = CV$. So a graph of the potential difference against time follows the same shape as Fig 66.

Fig 66
Charge versus time for a capacitor being charged

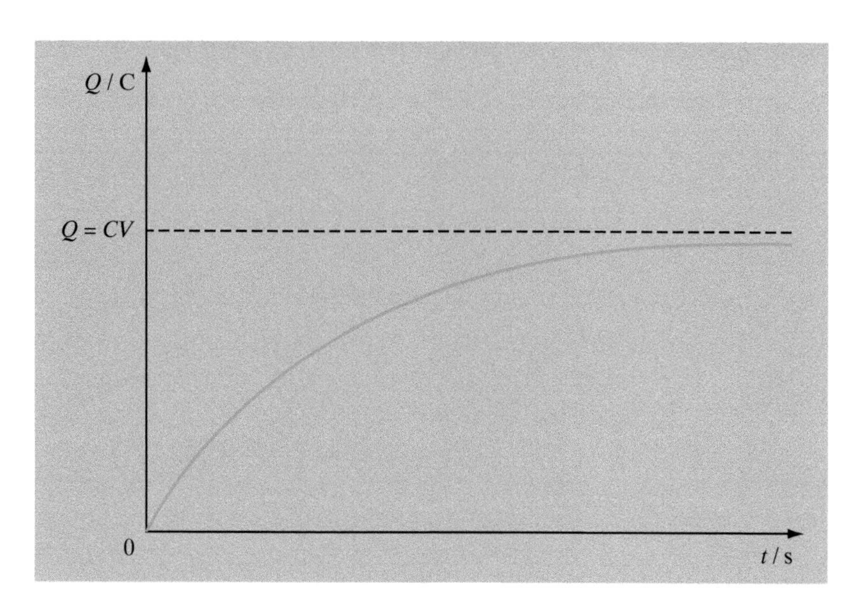

Essential Notes

Current is rate of change of charge, $\Delta Q/\Delta t$, so at any time the current is the gradient of this curve. The gradient of a tangent drawn at a point on this graph is equal to the rate of flow of charge at that instant in time, which is equal to the current.

Example

In the circuit of Fig 67, when the switch is closed the capacitor charges up and so the potential difference measured by the voltmeter goes up in proportion.

(a) The switch is closed at $t = 0$ s. Sketch a graph to show how the voltmeter reading changes over the next 15 seconds. Assume that the capacitor is initially uncharged.

(b) The experiment is repeated, but the voltmeter is connected across the resistor instead. How would a graph of the voltmeter readings look now?

(c) Explain the shape of the graphs in parts (a) and (b).

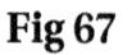

Fig 67

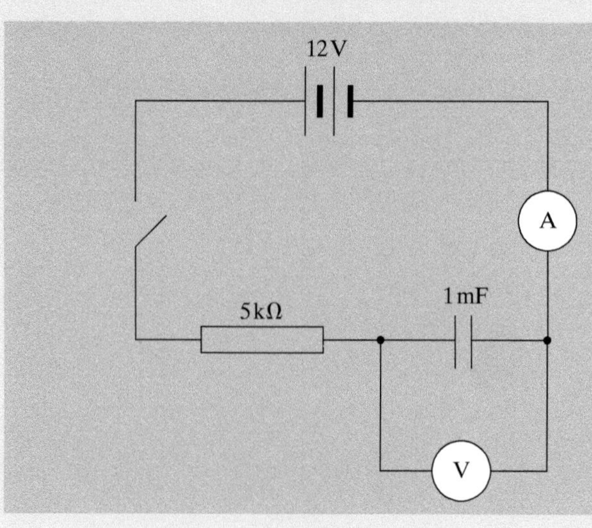

Answer

(a) and (b) See Fig 68.

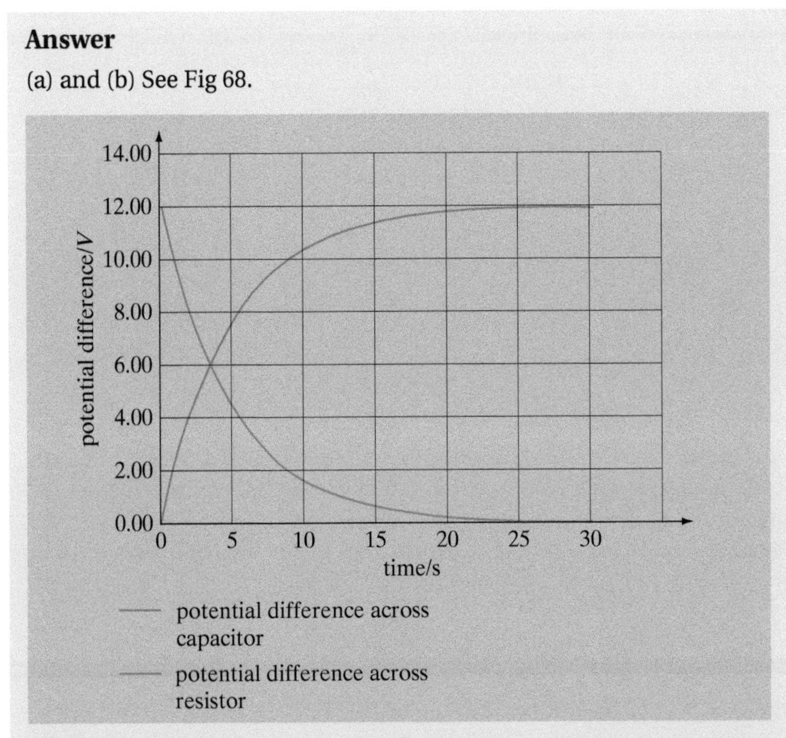

Fig 68 Potential difference versus time in the capacitor-charging circuit of Fig 67

(c) The potential difference across the capacitor is proportional to the charge stored, so it follows a similar shape to Fig 66. The potential difference is zero when the capacitor is uncharged and rises to a maximum value, equal to that of the battery.
The potential difference across the resistor is proportional to the current in the circuit, $V = IR$, and so it follows a similar shape to Fig 64. The potential difference is a maximum at $t = 0$ and approaches zero as time goes on.
At any time, the sum of the potential differences across the capacitor and the resistor must equal that of the battery, $E = V_{capacitor} + V_{resistor}$.

Charging a capacitor: calculations

When a capacitor is charged through a resistor (Fig 63), the current, I, decreases with time, t, according to the equation:

$$I = I_0 \, e^{-t/CR}$$

where I_0 is the initial current, when the circuit is first switched on. This is the equation of the curve in Fig 69.

The potential difference, V_R, across the resistor R, is given by the equation $V_R = IR$, so the change in potential difference across the resistor with time is given by:

$$V_R = V_0 \, e^{-t/CR},$$

or since the initial potential is E, the emf of the cell, $V_R = E \, e^{-t/CR}$

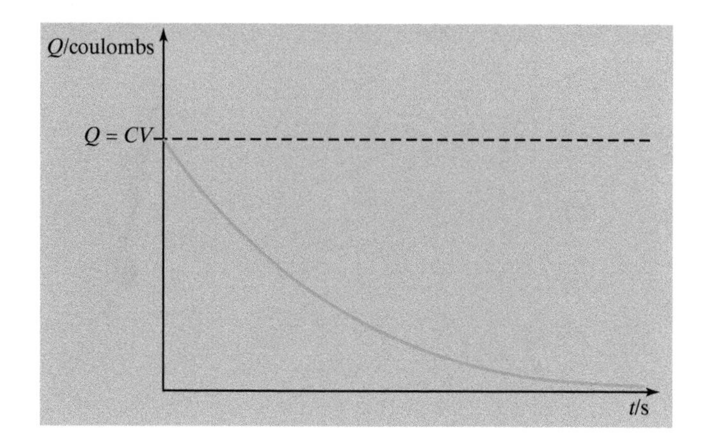

Fig 69
Charge versus time for a discharging capacitor. The graph of V versus time has the same shape.

Essential Notes

As noted in Fig 66, the current at any time is the gradient of the charge versus time graph. The exponential function has the unique property that the gradient at any point is proportional to the value of the function at that point. The current versus time graph therefore has the same shape as the charge (or potential difference) versus time graph.

The charge stored on the capacitor, Q, and the potential across it, V_C, are both initially zero, when the switch is first turned on. We can calculate their values at any time t using these equations:

$$V_C = E\,(1 - e^{-t\,/\,CR})$$

$$Q = Q_0\,(1 - e^{-t\,/\,CR}),$$

where Q_0 = maximum value of charge = CE

We use natural logarithms (logarithms to the base e) to solve these equations. For example, to find the time taken for the charging current to drop to half of its initial value:

$$I = I_0\,e^{-t}/RC$$

$$I/I_0 = e^{-t}/RC$$

Taking logarithms: $\ln\,(I/I_0) = \ln\,(e^{-t}/RC)$

$$\ln\,(I/I_0) = -t/RC$$

$$t = -(1/RC)\,\ln\,(I/I_0)$$

Since $I/I_0 = 0.5$ and $-\ln x = \ln\,(1/x)$

$$t = (1/RC) \times \ln 2 = 0.693/RC$$

Example

The circuit in Fig 70 is turned on at $t = 0$ s. (You can assume that the capacitor is completely uncharged at that time.)

(a) What will the current in the circuit be 10 s after the circuit has been turned on?

(b) How long before the capacitor is charged to half of its maximum value?

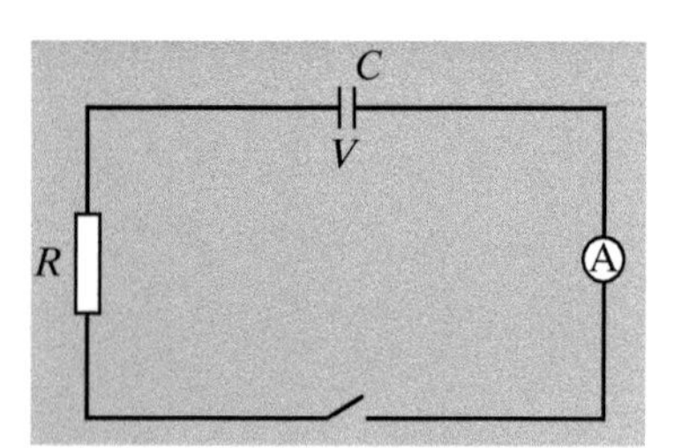

Fig 70 Circuit for discharging a capacitor

Answer

(a) Use $I = I_0\,e^{-t\,/\,CR} \rightarrow I = (12\,/\,5000)\,e^{-10\,/\,(5000 \times 0.001)} = 3.2 \times 10^{-4}\,A = 0.32$ mA.

(b) $Q = Q_0\,(1 - e^{-t\,/\,CR})$, so

$Q\,/\,Q_0 = (1 - e^{-t\,/\,CR})$

$Q\,/\,Q_0 - 1 = -e^{-t\,/\,CR}$

$(1 - Q\,/\,Q_0) = e^{-t\,/\,CR}$, then take natural logarithms of both sides:

$\ln_e (1 - Q / Q_0) = -t / CR$

$CR \times \ln (1 - Q / Q_0) = -t$

In general, the time for the charge to reach half of its maximum value is $CR \ln (½)$ or 0.693 CR.

In this case $0.001 \times 5000 \times \ln (1 - ½) = -t$

$$5 \ln (½) = -t$$

$\ln (0.5) = -0.693$, so $t = 3.47$ s

Discharging a capacitor

When a charged capacitor is placed in a circuit to be discharged, the initial discharging current depends on the potential across the capacitor, V, and the resistance of the circuit, R.

The initial current that flows is given by $I_0 = V/R$, and since $V = Q/C$, this can be written $I_0 = Q/CR$. As the charge flows off the capacitor, the voltage drops and the current gets less. The current, charge, and the potential difference across the capacitor all decrease exponentially with time.

The potential difference across the resistor is equal in magnitude, but opposite in polarity, to the potential difference across the capacitor.

At any time the sum of the potential differences across the capacitor and the resistor will be equal to zero.

$$V_C + V_R = 0$$

Example

A voltage sensor and a datalogger were used to measure the potential difference across a capacitor as it is discharged through a resistor.

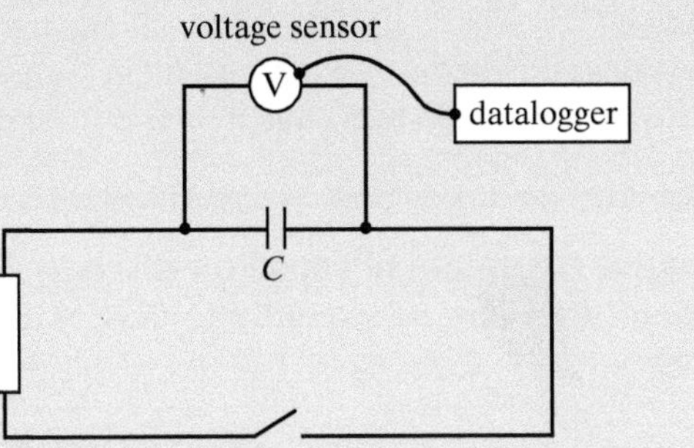

Fig 71

The following results were taken.

Time/s	0	1	2	3	4	5	6
Potential difference/V	10.00	8.19	6.70	5.49	4.49	3.68	3.01
Time/s	7	8	9	10	11	12	
Potential difference/V	2.47	2.02	1.65	1.35	1.11	0.91	

(a) Plot a suitable graph to find the time constant of the circuit.

(b) If the resistor had a value of 100 kΩ, calculate the value of the capacitor.

(c) The resistor and the capacitor are changed to values of 1 kΩ and 5 μF respectively. What sampling rate would you set the datalogger to record at? What would be a reasonable interval over which to record values?

(d) Explain how you could use a log–linear graph to find the value of the time constant.

Answers

(a)

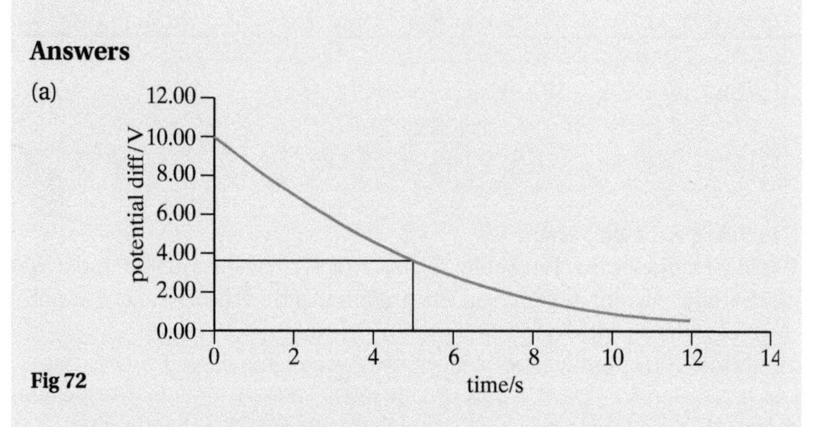

Fig 72

The voltage will drop to 1/e of its original value after 1 time constant. $V = 10/\mathrm{e} = 3.687$ V. Reading this value from the graph gives a time constant of 5.0 seconds.

(b) Time constant $= RC = 5.0$ s. So $C = 5.0/100\,000 = 5 \times 10^{-5}$ F $= 50$ μF.

(c) The new time constant is $1 \times 10^3\ \Omega \times 5\ \mu\text{F} = 5 \times 10^{-3}$ s $= 5$ ms.

The sampling rate needs to be significantly faster than this, say a minimum of once every 0.5 ms, which is a rate of 2000 Hz. After five time constants the value will have dropped to less than 0.67% of the original voltage. So a logging time of 25 ms should be sufficient.

(d) $V = V_0\,\mathrm{e}^{-t/RC}$, so $\ln V = \ln V_0 - t/RC$. Compare this with the equation of a straight line, $y = mx + c$. Plot $\ln V$ on the y-axis and t on the x-axis, which should be a straight line with a gradient of $-1/RC$. So $RC = -(1/\text{gradient})$.

Essential Notes

The graph has $\log V$ on the y-axis, but the time scale is linear on the x-axis. This is referred to as a **log–linear graph**. (See Guidance on logarithms and logarithmic graphs, page 127.)

Quantitative treatment of capacitor discharge

The equation describing the exponential decay of a discharging capacitor is

$$Q = Q_0\mathrm{e}^{-t/RC}$$

where Q is the charge at time t, Q_0 is the initial charge, C is the value of the capacitance and R is the resistance of the discharging circuit. When $t = RC$, this equation becomes

$$Q = Q_0\mathrm{e}^{-RC/RC}$$

$$\frac{Q}{Q_0} = \mathrm{e}^{-1} = 0.37 \quad \text{(to 2 s.f.)}$$

Using the equation for the charge during discharge, we can find the time taken to drop to a certain charge (or potential):

$$Q = Q_0\mathrm{e}^{-t/RC}$$

Taking natural logs,

$$\ln\left(\frac{Q}{Q_0}\right) = -\frac{t}{RC}$$

So

$$t = -RC \ln\left(\frac{Q}{Q_0}\right)$$

Essential Notes

Don't worry about the negative sign in front of the time. The log of a fraction is always negative so the negative signs will cancel, giving a positive time.

Example

A 10 μF capacitor is charged to a potential difference of 12 V and then discharged through a 100 kΩ resistor. How long would it be before the potential difference across the capacitor dropped to 1 V?

Answer

$Q = Q_0 e^{-t/RC}$, but since $Q = CV$,

$V = V_0 e^{-t/RC}$

Putting in the potential difference values,

$1 = 12 \times e^{-t/RC}$

Since $RC = 10 \times 10^{-6}\,\text{F} \times 100 \times 10^3\,\Omega = 1\,\text{s}$,

$1 = 12 \times e^{-t}$ and so $e^t = 12$

Taking natural logs,

$t = \ln 12$

giving $t = 2.48\,\text{s}$.

3.7.5 Magnetic fields

3.7.5.1 Magnetic flux density

A magnetic field can be created by a permanent magnet or by an electric current. A magnetic field is a region where another permanent magnet experiences a force. The field lines around a magnet show the direction of the force on a north pole.

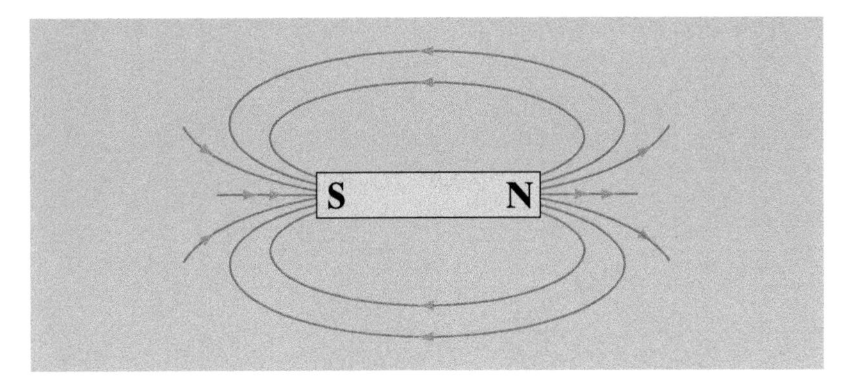

Fig 73
The field lines around a bar magnet

Essential Notes

Magnetic fields have no effect on stationary charges.

Magnetic fields affect moving charges. Since a flow of current in a wire is due to a stream of moving charges, any current-carrying conductor in a magnetic field experiences a force (Fig 74). This effect, known as the **motor effect**, arises due to the interaction of the magnetic field around the moving charges with the external field.

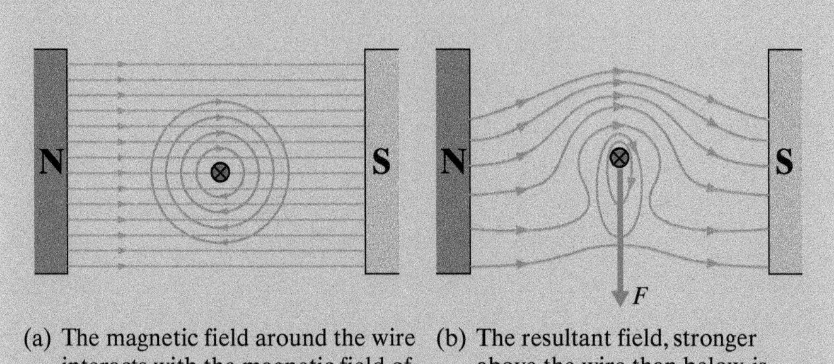

(a) The magnetic field around the wire interacts with the magnetic field of the permanent magnet

(b) The resultant field, stronger above the wire than below, is sometimes called a 'catapult field'

Fig 74
Catapult field around a current-carrying wire, giving rise to a force

Essential Notes

A cross, ×, is used to show a current, or a magnetic field line, pointing away from you into the plane of the paper. A dot is used to show a current, or a magnetic field line, pointing towards you out of the plane of the paper.

The direction of the force is at right angles to both the direction of the current and the direction of the field. Fleming's left-hand rule (Fig 75) is a way of remembering the relative directions of the magnetic field, the electric current and the resulting force.

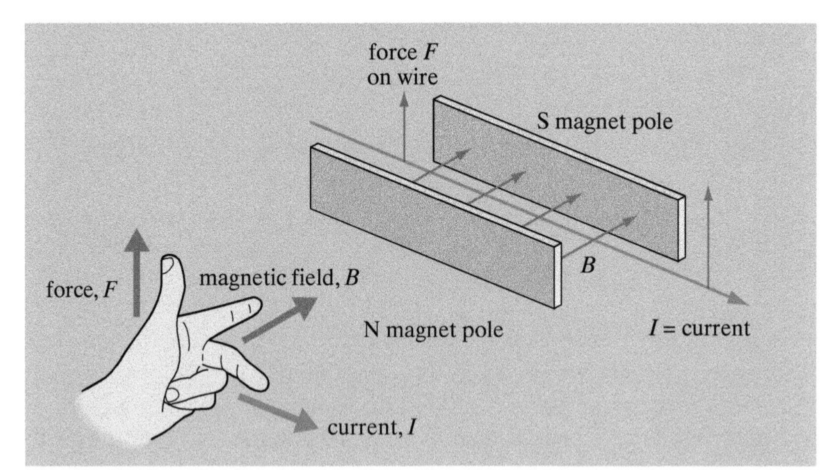

Fig 75
Fleming's left-hand rule gives the direction of the force.

The first finger must point in the direction of the field, from north towards south. The second finger must point in the direction of conventional current, from + to −. The thumb will then indicate the direction of the force on the wire.

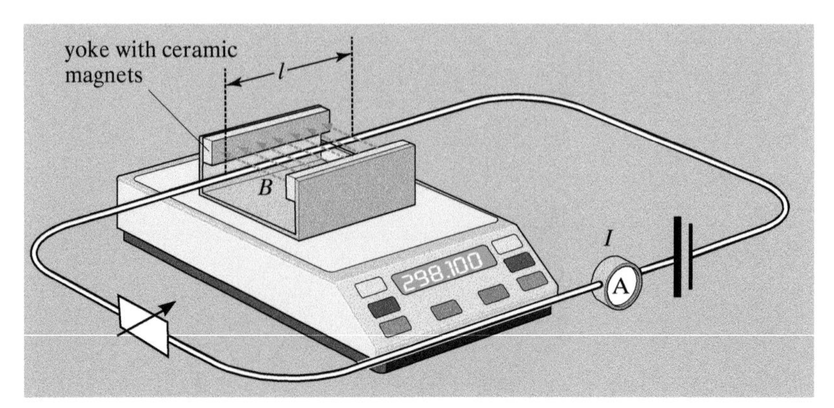

Fig 76
Apparatus to investigate the size of force on a current-carrying wire

The size of the force can be investigated with the apparatus in Fig 76. Experiments like this show that the size of the force, F, dependent on:

- the strength of the magnetic field;
- the size of the electric current, I;
- the length of the wire, l, that is in the magnetic field.

In SI units the force F is given by

$$F = BIl$$

Where B is called the **magnetic flux density** which has SI units of **tesla**,T. We use the equation $F = BIl$ to define the tesla. Since $B = F/Il$, the magnetic flux density is the force per unit length on a wire carrying a current of one ampere.

Essential Notes

If the current and the wire are not at right angles, the force becomes $F = BIl \sin \theta$ where θ is the angle between the current and the field.

Definition

One tesla is the magnetic flux density that causes a force of 1 N to be exerted on every 1 m length of a wire carrying 1 A of current in a direction that is perpendicular to the field.

The tesla is quite a large unit; a typical permanent magnet would have a flux density close to the poles of around 100 mT, whilst the strongest electromagnets in the world create maximum flux densities of only a few tens of tesla.

3.7.5.2 Moving charges in a magnetic field

The force on a current-carrying wire in a magnetic field is given by $F = BIl$. This equation can be adapted to apply to a charge Q that is moving in a magnetic field at a velocity v.

$$I = \frac{Q}{t}, \text{ so } F = B\left(\frac{Q}{t}\right)l$$

This can be written as

$$F = BQ\left(\frac{l}{t}\right)$$

and since $\frac{l}{t}$ = velocity, v, the equation becomes

$$F = BQv$$

Definition

The force on a charged particle moving through a magnetic field is F = BQv, where Q is the charge of the particle and v is its velocity perpendicular to the magnetic field.

Notes

Use Fleming's left-hand rule (page 80) to find the direction of the force. The second finger (current) points in the direction that a positive charge is moving.

Essential Notes

Remember to reverse the direction of your second finger if it is a negative particle, such as an electron.

Because the force is always at right angles to the particle's velocity, a charged particle moving in a magnetic field has a circular path. The magnitude of the particle's velocity does not change, but its direction is constantly changing.

The magnetic force, BQv, causes a centripetal acceleration. For a particle of mass m,

$$BQv = \frac{mv^2}{r}$$

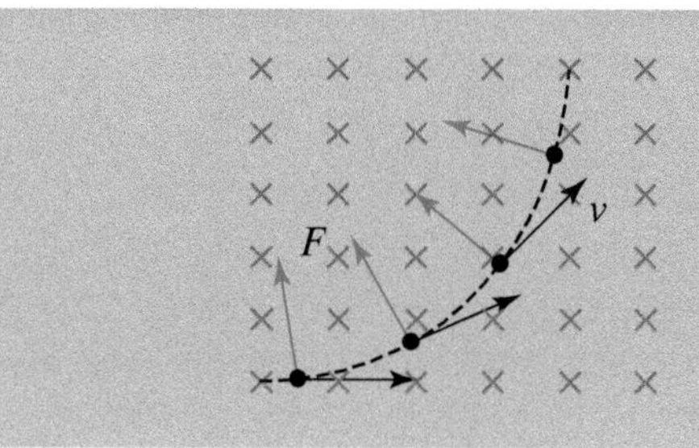

Fig 77
A moving charged particle follows a circular track in a magnetic field because the force on it is always at 90° to its velocity.

The radius of the path is therefore given by

$$r = \frac{mv}{BQ}$$

Magnetic fields are used to deflect charged particles in particle accelerators like the cyclotron. In a cyclotron a source of charged particles is placed in the centre of two 'D'-shaped regions. A magnetic field curves the particles' path into a semicircle in each D.

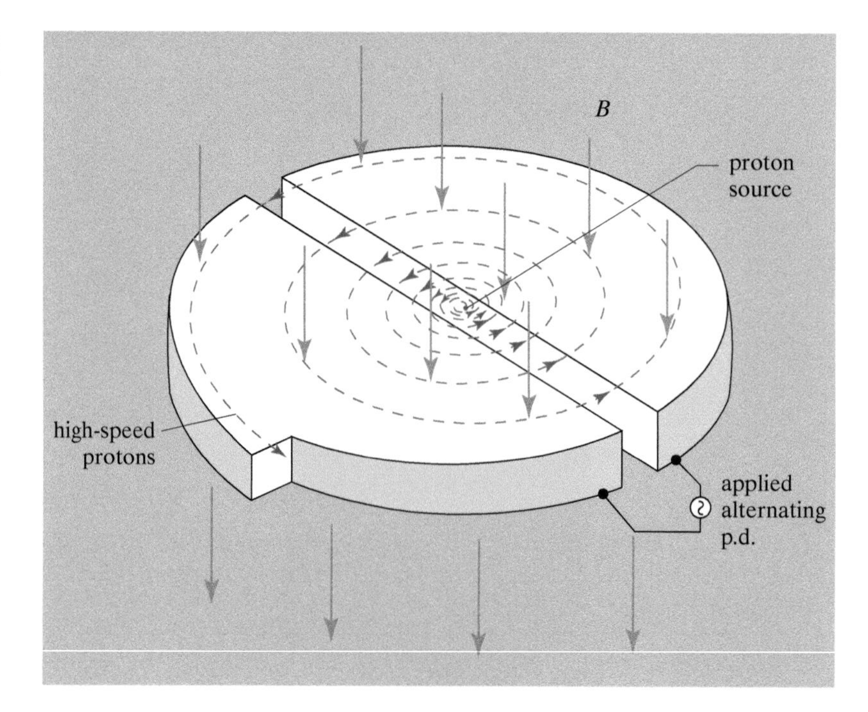

Fig 78
A cyclotron

An alternating potential difference accelerates the particles as they pass across the gap between the Ds. For the cyclotron to work, the time taken for a particle to travel one half-circle must match the time taken for the polarity of the a.c. to reverse, so that the particles arrive at the gap at just the right time to be accelerated. The time taken, t, for a particle to travel a whole circle is

$$t = \frac{2\pi r}{v} = 2\pi r \times \frac{m}{BQr} = \frac{2\pi m}{BQ}$$

So the travel time does not depend on the radius of the path. Every time a particle arrives at the gap it will be accelerated by the electric field. The particles travel in a series of semicircular paths with gradually increasing radius. They move at higher speeds on each turn, so that the cyclotron's limiting factor is its size: bigger cyclotrons can produce faster particles.

Essential Notes

Increasing the size of a cyclotron only works up to a certain limit. As the particles approach the speed of light their mass increases and the orbital time changes. The relativistic increase in mass is covered in the optional unit Turning points in physics.

3.7.5.3 Magnetic flux and flux linkage

The magnetic flux density can be thought of as the number of flux lines that pass through a given area. In a strong field the field lines are close together and the flux density is high.

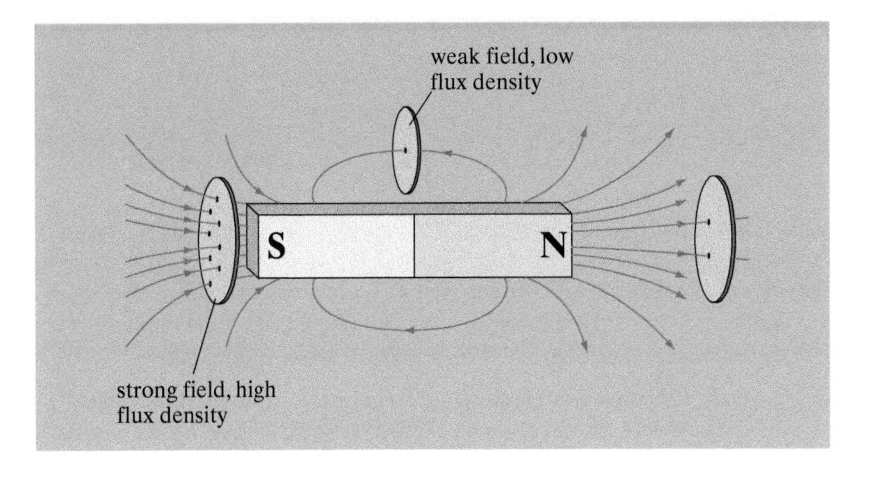

Fig 79
Flux density and field lines

Provided that the area is at right angles to the field lines, the total **flux**, Φ, that passes through a given area, A, is given by

$$\text{total flux} = \text{flux density} \times \text{area}$$

or

$$\Phi = BA$$

Essential Notes

Since flux density is the flux per unit area, one tesla is equal to 1 weber per square metre.

Flux is measured in units of **webers**, Wb.

Sometimes, especially when we arc considering electromagnetic induction (see below), it is useful to calculate the total flux through a coil of wire. If the coil has several turns of wire, the flux through the whole coil is the sum of the flux through each individual turn. This is referred to as the **flux linkage** through the coil. For a coil of N turns the flux linkage is $N\Phi$.

Fig 80
Flux linkage is $BAN \cos\theta$.

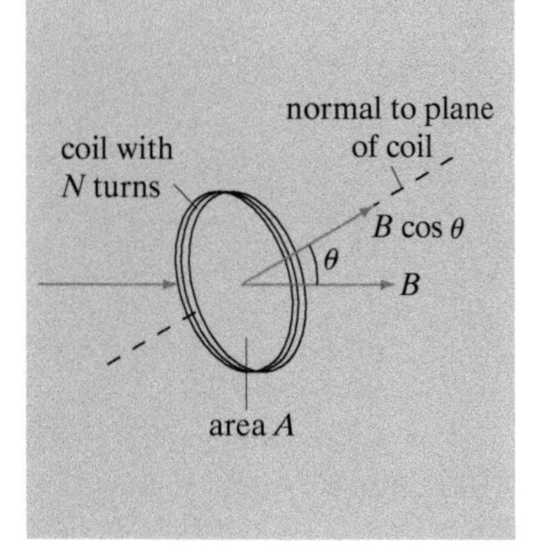

The amount of flux linkage through a coil depends on the orientation of the coil with respect to the magnetic field. If θ is the angle between the normal to the plane of the coil and the magnetic field (Fig 80), the total flux Φ through the rectangular coil is $BA \cos\theta$. If there are N turns on the coil, the total flux linkage is

$$N\Phi = BAN \cos\theta$$

3.7.5.4 Electromagnetic induction

When a bar magnet is moved relative to a coil of wire connected in a circuit, an electric current is made to flow in the coil. The potential difference that causes this current is referred to as an **induced e.m.f.** The direction of the induced e.m.f. depends on the direction of motion.

Fig 81
Simple electromagnetic induction apparatus

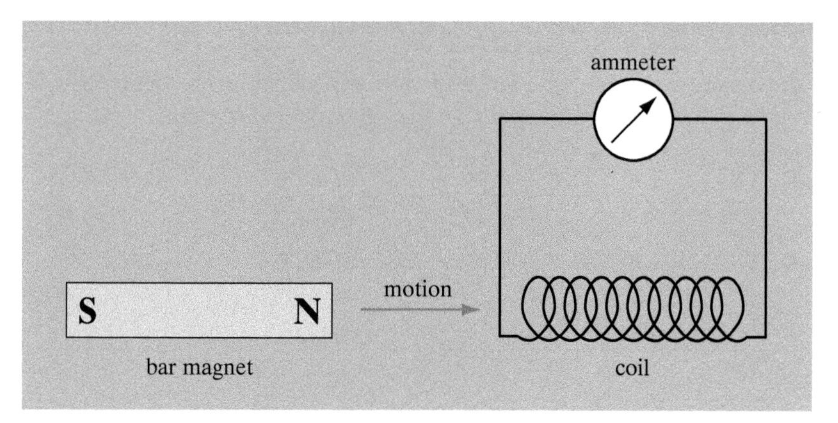

As the north pole of the magnet is pushed towards the coil, the induced e.m.f. will cause a current to flow in the coil (Fig 81). This current will create a magnetic field around the coil. There are two possibilities:

1 The end of the coil nearest the magnet could become a south magnetic pole. This would attract the bar magnet, thereby increasing the rate at which flux changes through the coil. This would increase the induced e.m.f. and current and make the magnet stronger, which would attract the magnet with a greater force. This sequence of events would continue and would increase the kinetic energy of the magnet at the same time as increasing the electrical energy generated. This would contravene the conservation of energy and so it does not happen.

2 The end of the coil nearest the magnet could become a north pole. This would repel the bar magnet. Indeed the quicker you tried to push the magnet into the coil, the larger the induced e.m.f., and the larger the current and the greater the force opposing the motion.

Definition

Lenz's law states that the direction of the induced e.m.f. always acts so as to oppose the change that is causing the e.m.f.

Faraday's law

We can use the same apparatus (Fig 81) to investigate the factors that affect the magnitude of the induced e.m.f.

The size of the e.m.f. that is induced in the coil depends on:

- the relative speed of the coil and the magnet: the faster the movement, the larger the e.m.f.;
- the number of turns on the coil;
- the area of the coil;
- the strength of the magnetic field.

Faraday combined these effects into a single statement.

Definition

***Faraday's law** of electromagnetic induction states that the magnitude of the induced e.m.f., ε, is proportional to the rate of change of **magnetic flux linkage** through the coil.*

For a coil of N turns, this can be written as

$$\varepsilon = \frac{N\Delta\Phi}{\Delta t}$$

Example

A small circular coil of radius 2 cm is placed close to a bar magnet so that its plane is perpendicular to the field lines. The magnetic flux density is 40 mT. The coil, which has 20 turns, is turned so that it is edge-on to the field. If this movement takes 0.1 s, calculate the induced e.m.f.

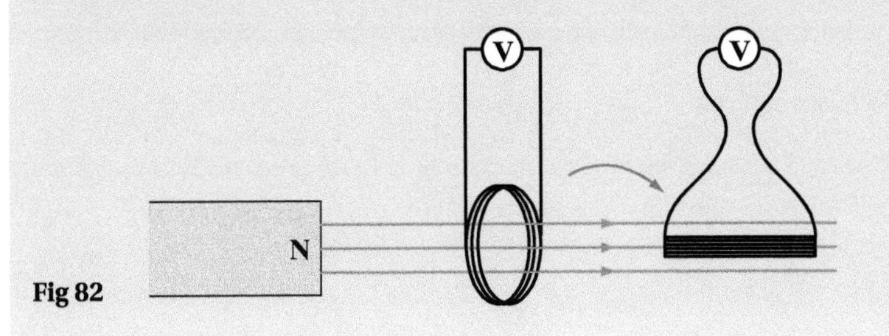

Fig 82

Answer

Initially, the total flux linkage through the coil is:

$N\Phi = BAN = 40 \times 10^{-3} \times \pi(0.02)^2 \times 20 = 1.01 \times 10^{-3}$ Wb turn

When the coil is turned the flux linkage drops to zero, so

$N\,\Delta\Phi = 1.01 \times 10^{-3}$ Wb turn

So

$$\varepsilon = N\,\Delta\Phi/\Delta t = \frac{1.01 \times 10^{-3}}{0.1} = 0.01\text{ V}$$

Faraday's law applies more generally to any situation where there is relative motion between a conductor and a magnetic field. Another way of thinking about the law is to say that the induced e.m.f. is proportional to the rate at which field lines are cut through.

An aircraft flying horizontally cuts through the vertical component of the Earth's magnetic field. An e.m.f. is generated between the tips of its wings. To calculate the e.m.f. we need to calculate the flux cut through per second.

Area travelled through in 1 second A = length of wings × velocity = $l \times v$

Flux cut through in 1 second gives the induced e.m.f.:

$$\varepsilon = BA = Blv$$

The vertical component of the Earth's magnetic field is 4×10^{-5} T so if the wingspan of the aircraft is 40 m and the aircraft's speed is 130 m s^{-1}, the induced e.m.f. will be

$$\varepsilon = 4 \times 10^{-5} \times 130 \times 40 = 0.21 \text{ V}$$

The e.m.f. induced in a rotating coil

A simple generator can be constructed from a coil rotating in a magnetic field, Fig 83.

Fig 83
A simple generator

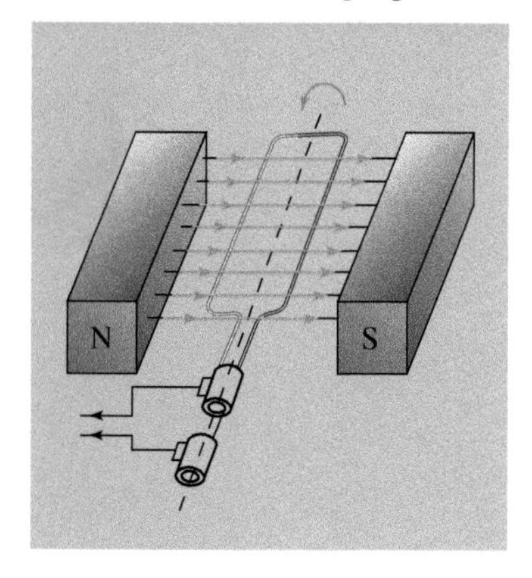

The induced e.m.f. in the coil is equal to the rate of change of magnetic flux linkage, according to Faraday's law. Therefore, whenever the coil is turning and cutting through magnetic field lines, an e.m.f. will be induced. The greater the rate of cutting field lines, the greater the e.m.f.

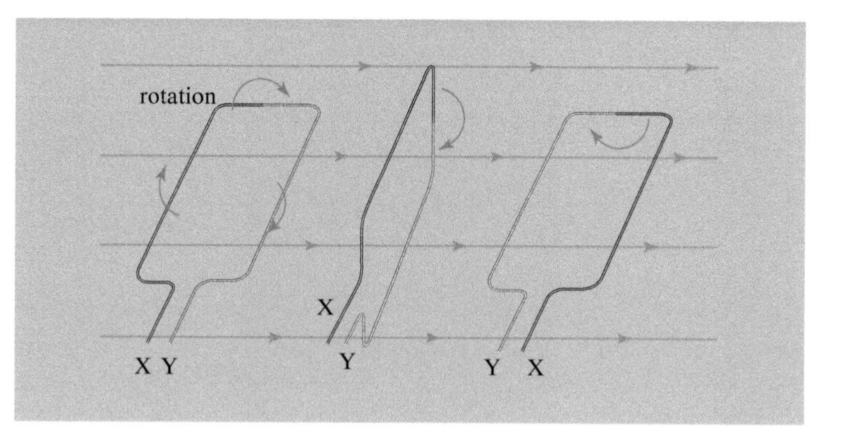

Fig 84
Changing flux linkage through the rotating coil

When the plane of the coil is parallel to the field lines, the flux linkage is zero but the rate of change of flux linkage is maximum (Fig 84). When the plane of the coil is perpendicular to the field, the flux linkage is maximum but the rate of change of flux linkage is zero. When the plane of the coil is again parallel to the field, the rate of flux linkage is again maximum, but the sides of the coil are moving through the field in the opposite direction (in Fig 84, side X is moving down instead of up). Thus an alternating e.m.f. is induced, as shown in Fig 85.

The flux linkage through the coil is $BAN \cos\theta$ (see page 84). The e.m.f. induced across the coil is given by the rate of change of flux linkage = $\Delta(BAN \cos\theta)/\Delta t$, which is the gradient of graph 1 in Fig 85. Graph 2 shows how the gradient of graph 1 changes with time. This is the induced e.m.f; it varies sinusoidally with time.

If the coil rotates with angular frequency $(\Delta\theta/\Delta t) = \omega$ rad s^{-1}, the induced e.m.f. at time t is given by

$$\varepsilon = BAN\omega \sin \omega t$$

This varies between 0 and $BAN\omega$.

Essential Notes

We can derive the expression for the induced e.m.f. using calculus. Differentiating the expression for flux linkage $BAN \cos \theta$ with respect to time to obtain the rate of change of flux linkage, with $\theta = \omega t$, gives $-BAN\omega \sin \omega t$ for the induced e.m.f. The negative sign shows that the e.m.f. is in a direction that opposes the rotation, in accordance with Lenz's law.

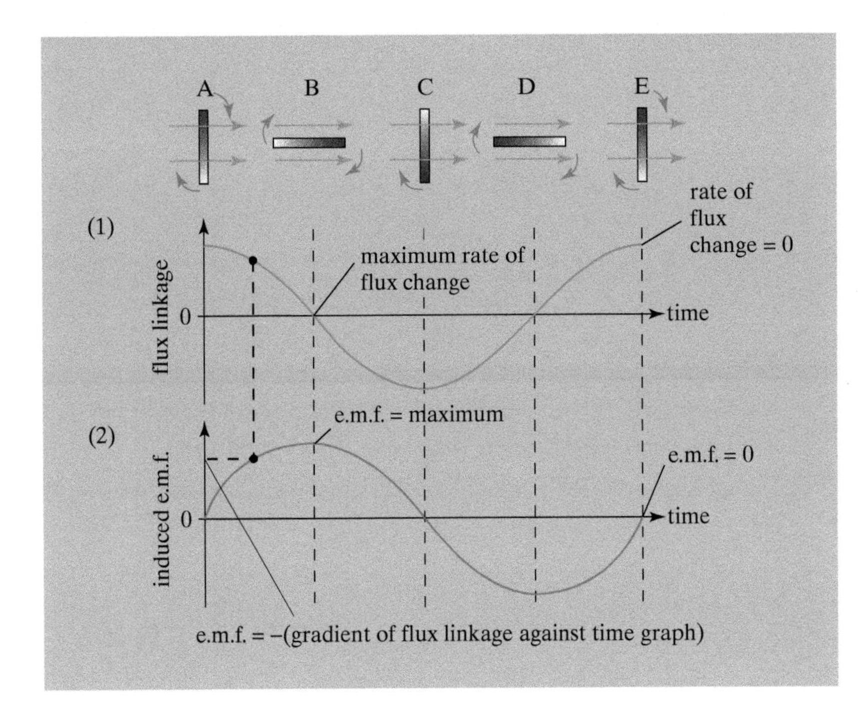

Fig 85
Induction of an alternating e.m.f.

A: The coil is face-on to the field and flux linkage is maximum
B: The coil is edge-on to the field and flux linkage is zero
C: The coil is face-on to the field but moving in the opposite direction to A
D: The coil is edge-on to the field and flux linkage is zero
E: The coil is face-on to the field and flux linkage is maximum

Example

A bicycle dynamo is made from a circular coil of 100 turns of wire which rotates between two magnets where the flux density is 40 mT. The radius of the coil is 1.5 cm.

(a) If the coil rotates at 1500 rpm, calculate the peak voltage produced.

(b) The e.m.f. generated by the dynamo is shown on an oscilloscope. Sketch the trace you would expect to see if the bicycle travelled twice as fast.

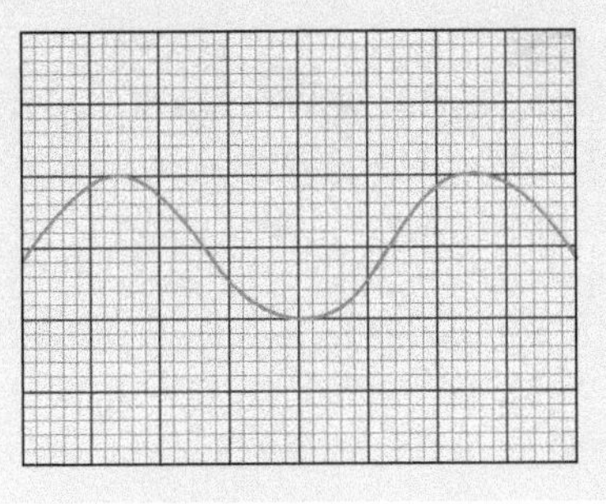

Answer

(a) Peak e.m.f. = $BAN\omega$

$A = \pi r^2 = 3.14 \times (1.5 \times 10^{-2})^2 = 7.07 \times 10^{-4}\ \text{m}^2$

$\omega = 2\pi f = 2\pi \times 1500/60 = 157\ \text{rad s}^{-1}$

Peak e.m.f. $= 40 \times 10^{-3} \times 7.07 \times 10^{-4} \times 100 \times 157 = 0.44\ \text{V}$

(b) The peak voltage (amplitude) of the signal would double. The frequency of the signal would also double.

Notes

From Faraday's law, since the rate at which the flux linkage changes will double, the peak induced e.m.f. will also double. The dynamo will turn twice as fast so the coil will go through twice as many cycles per second (see Fig 85). This will double the frequency of the a.c.

Alternating currents

- The term **a.c.** can refer to **alternating current** and also alternating p.d.
- The term **d.c.** can refer to **direct current** and also direct p.d.
- a.c. can be represented as a sinusoidal graph varying between zero and a maximum value, either side of zero, in opposite directions (Fig 86).

Fig 86
Alternating p.d. and current

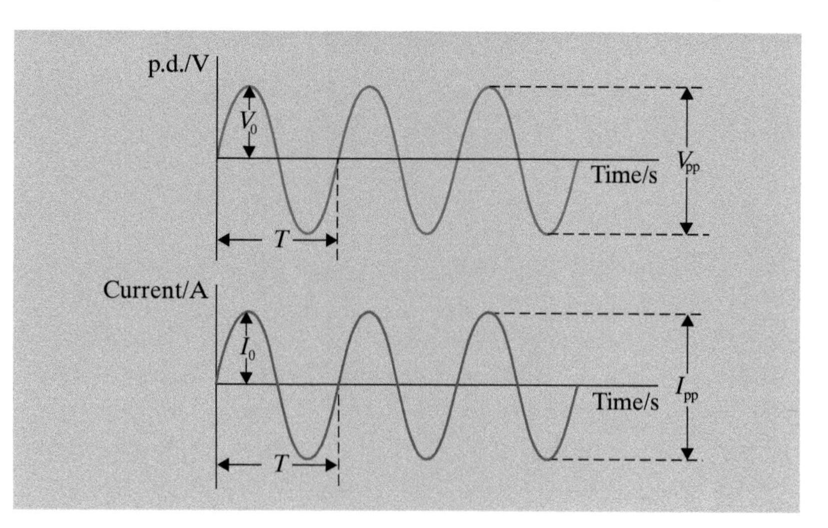

Essential Notes

V_{pp} is peak-to-peak p.d.

I_{pp} is peak-to-peak current.

Several quantities can be calculated from graphs such as those in Fig 86, including root mean square value, peak value, peak-to-peak value, time period and frequency.

The **root mean square value (r.m.s.)** – since the a.c. signal is continually changing in value it is impossible to assign a fixed value over a number of cycles (the mean would be zero). The r.m.s. current produces the same heating effect in a resistor as the equivalent d.c.; i.e. I_{rms} produces the same heating effects as $I_{d.c.}$

The **peak value** of an a.c. signal is the maximum displacement from the zero line in either direction and is labelled either I_0 or V_0 respectively on the graph.

To convert peak values to r.m.s. values the following equations can be used:

$$I_{rms} = \frac{I_0}{\sqrt{2}} \qquad V_{rms} = \frac{V_0}{\sqrt{2}}$$

The **peak-to-peak value** of an a.c. signal is the maximum displacement across both directions and is labelled either I_{pp} or V_{pp} respectively on the graph.

The **time period** (T) of an a.c. signal is the time taken for one complete cycle. The unit is seconds or, more usually, ms or μs.

The **frequency** (f) of an a.c. signal is the number of complete cycles per second. The unit is hertz (Hz).

Period and frequency are linked by the equation:

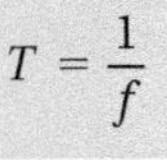

$$T = \frac{1}{f}$$

Essential Notes

In a practical situation it is better to measure V_{pp} and divide by 2 in order to obtain a value for V_0 since there is less error in measuring a longer length.

Oscilloscope

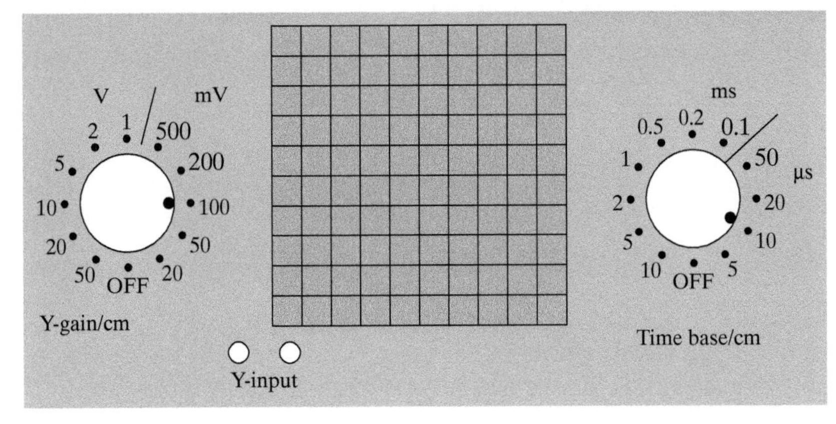

Fig 87
A typical oscilloscope

An **oscilloscope** is a device for displaying waveforms. It can be used to:

- measure a.c. and d.c. voltages
- measure small time intervals
- measure frequencies of alternating currents and voltages.

Interpreting the oscilloscope screen

Oscilloscopes are used to investigate how a potential difference changes with time. The potential difference is applied to the Y-input (see Fig 87). The screen will show a graph of potential difference (y-axis) against time (x-axis). The controls on the oscilloscope allow us to adjust the scales on the graph.

- Y-gain – This controls the scale on the y-axis (i.e. the sensitivity of the oscilloscope). The control is marked in volts per cm or millivolts per cm.
- Time base – This controls the time scale on the x-axis. The control is marked in milliseconds, ms, or microseconds, μs.

Using the oscilloscope

When the oscilloscope is first turned on, the time base and the Y-gain should both be turned to OFF. The screen will show a spot in the centre. If the time base is turned on, the spot will move left to right horizontally across the screen. As shorter times are selected the spot moves more quickly, soon appearing as a horizontal line (see Fig 88).

Measuring a d.c. signal

To measure a constant d.c. voltage (such as that from a battery or cell) applied to the Y-input, the Y-gain is turned on. If the time base is turned off, the spot will rise. If the time base is turned on, the horizontal line will rise. The Y-gain should be set at the most sensitive setting possible, i.e. one that keeps the line on the screen. For example, if the Y-gain was set at 0.5 V/cm and the line moved up 3 cm, the d.c. voltage must be $0.5 \times 3 = 1.5$ V.

Fig 88
Oscilloscope screen with an external p.d. applied

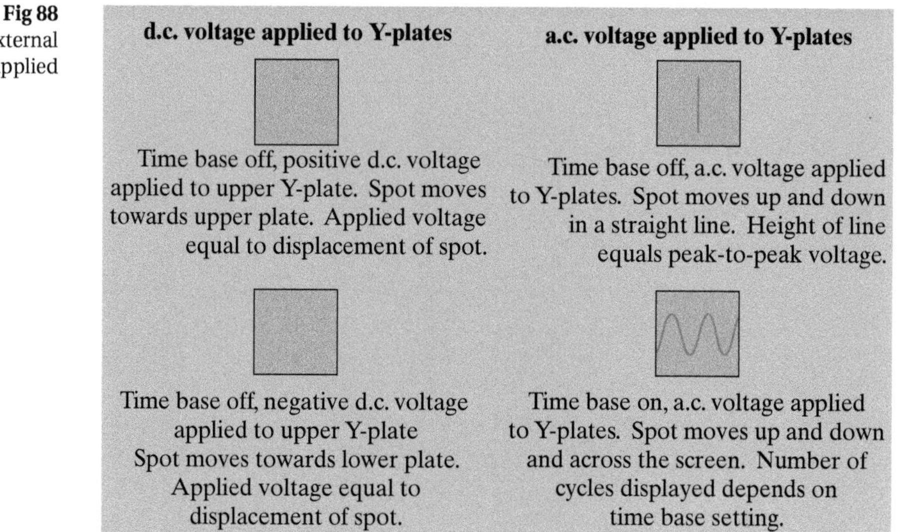

Measuring an a.c. signal

To measure an a.c. signal, such as that from a signal generator or the mains supply, the time base and Y-gain are both turned on and adjusted until a steady signal is seen on the screen. We can calculate the frequency and peak voltage of the a.c. signal from the screen display.

Suppose the screen appeared as in Fig 89, with the time base set at 2 ms/cm and the Y-gain set at 0.2 mV/cm.

Frequency: to get a **precise** calculation we use as much of the screen as possible. Six cycles takes up about 9.5 cm on the screen, but each cm represents 2 ms, so the time for one cycle is $(9.5 \times 2)/6 = 3.2$ ms. As frequency, $f = 1/T$:

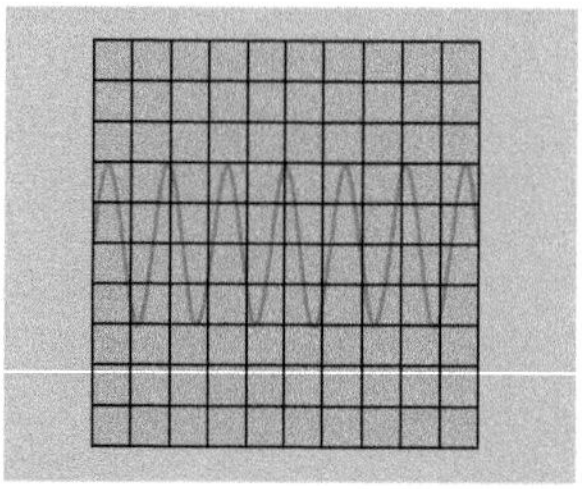

$f = 1 / 3.2$ ms $= 1 / 0.0032 = 312.5$ Hz. But the readings are only given to 2 s.f., so frequency = 310 Hz.

Fig 89
Measuring period and frequency of a.c.

Peak voltage: the screen should be adjusted so that the vertical displacement is as large as possible. Using Fig 89, the peak-to-peak distance is 4 cm, but each cm represents 0.2 mV, so the peak-to-peak voltage = 4 × 0.2 = 0.8 mV. The peak voltage is half of this, so V0 = 0.4 mV.

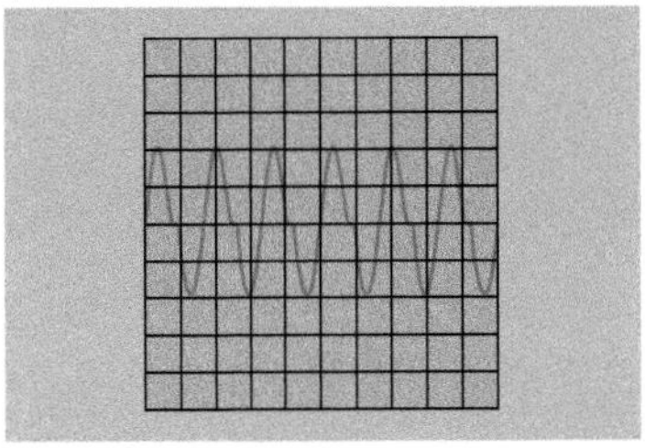

Example

Use the oscilloscope trace in Fig 90 to calculate the heartbeat rate per minute. The time base setting is 0.4 s/division.

Answer

Time base setting = 0.4 s/division

There are 6 heartbeats in 10 divisions

Length of one beat = 1.67 divisions

Period = 1.67 × 0.4 = 0.67 s

Frequency $= \frac{1}{0.67} = 1.5$ Hz or 90 beats per min

Fig 90
Oscilloscope trace of a heartbeat

Example

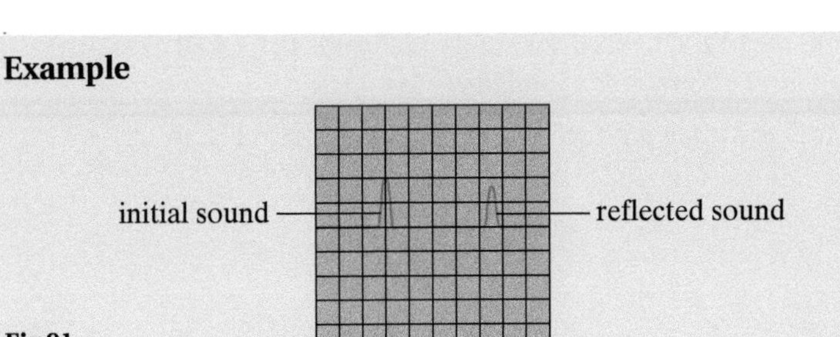

Fig 91

A sound and its reflection travelling through an iron bar and back are picked up by a microphone and displayed on an oscilloscope (Fig 91). If the time base is set at 2 ms/division, how long does the sound take to travel down the bar and back?

Answer

Time base = 2 ms/division

There are 4.5 divisions between pulses.

Total time elapsed $= 2 \times 10^{-3} \times 4.5 = 9$ ms

Example

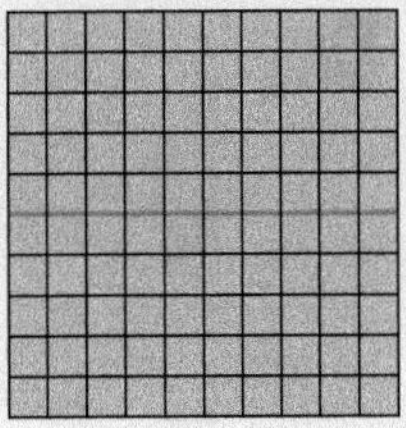

Fig 92

Fig 92 shows an oscilloscope screen with a line running centrally across the screen. The time base is set to 2 ms/division and the Y-gain setting is 2 V/division.

Draw the trace seen on the screen when a 6 V_{rms} a.c. signal of frequency 200 Hz is applied to the Y-inputs.

Answer

Calculate the number of cycles on screen.

Frequency = 200 Hz

$$\text{Period} = \frac{1}{f} = \frac{1}{200} = 5 \times 10^{-3}\,\text{s or } 5\,\text{ms}$$

Time base is set at 2 ms/division and there are ten divisions on the screen. Therefore there will be 4 cycles on the screen.

Calculating the peak voltage.

$$V_{peak} = V_{rms} \times \sqrt{2} = 6 \times \sqrt{2} = 8.5\,\text{V}$$

Y-gain is set at 2 V/division so amplitude needs to be 4.25 divisions either side of central horizontal line.

Therefore the trace produced would be as shown in Fig 93.

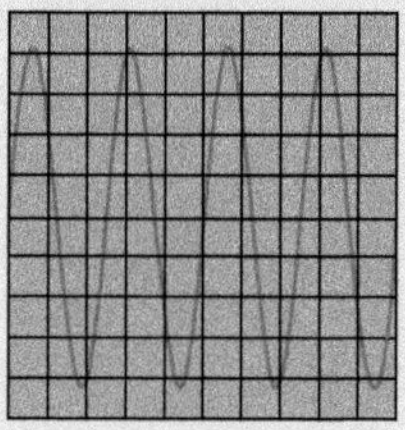

Fig 93

Example

An oscilloscope is connected to the terminals of a 12 volt car battery. The time base is turned off.

(a) Assume the oscilloscope is similar to the one in Fig 87. What setting would you use for the Y-gain?

(b) Describe and explain what you would expect to see on the screen.

(c) Describe and explain what the screen would show if the time base is now turned on, using a setting of 1 ms/cm.

Answer

(a) 5 V/cm would give a deflection of almost 2.5 cm. (2 V/ cm would give a deflection of 6 cm – off the screen!)

(b) The spot would deflect upwards by 2.5 cm.

(c) The spot would become a line.

3.7.5.6 The operation of a transformer

A **transformer** is used to change the voltage of an a.c. signal. It is made from two coils which are usually wound on the same piece of ferromagnetic material (such as soft iron).

Fig 94
The principle of a transformer

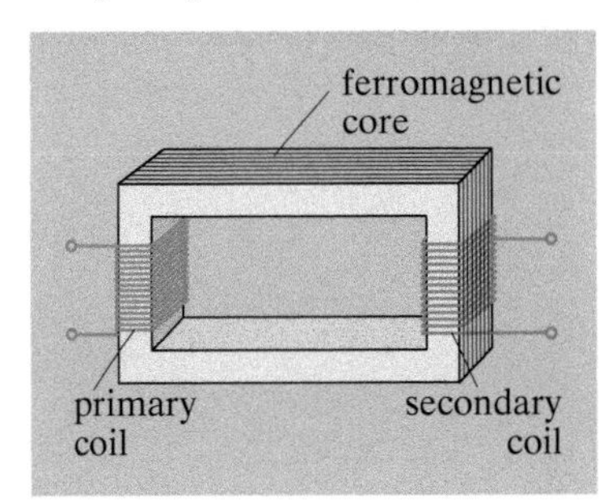

The primary coil is connected to an a.c. source, so there is an alternating current in the primary coil. The primary coil acts as an electromagnet which varies in strength as the current through it varies. The magnetic flux in the core therefore changes, changing the flux linkage through the secondary coil. This induces an alternating e.m.f. in the secondary coil.

If there are fewer turns on the secondary than on the primary, the output e.m.f. is smaller than the input voltage, and the transformer is said to be a step-down transformer. If there are more turns on the secondary than on the primary, the output voltage is greater than the input voltage, and the transformer is said to be a step-up transformer.

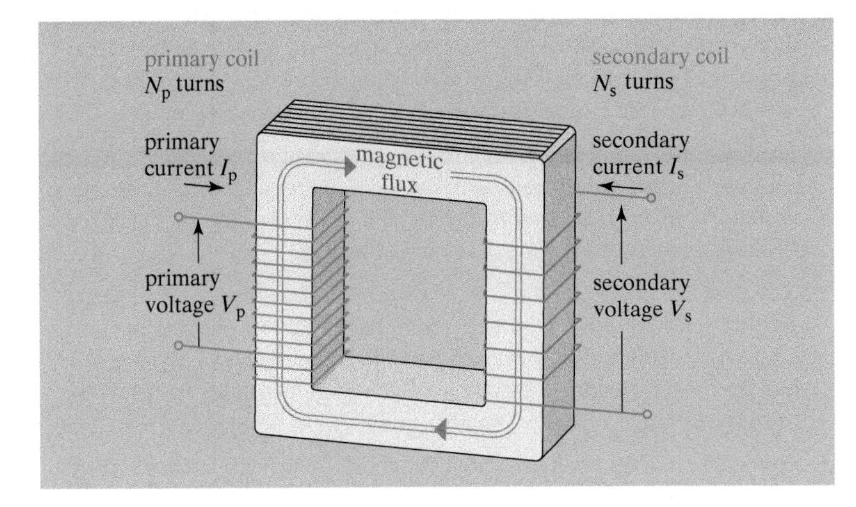

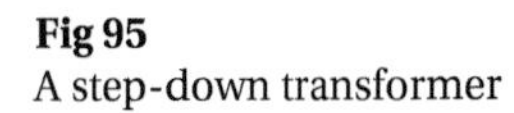
Fig 95
A step-down transformer

The ratio of the voltages across the primary and secondary coils is equal to the ratio of the number of turns.

Definition

The transformer equation is $\frac{N_s}{N_p} = \frac{V_s}{V_p}$

A step-up transformer increases the potential difference, so the output voltage is greater than the input voltage. But the power output cannot be greater than the power input, as this would contravene the principle of energy conservation. The most we could expect, for a 'perfect' transformer, is that the output power would equal the input power. That would mean that no energy is dissipated in the transformer, in other words it would be 100% efficient. Since power $P = IV$,

$$I_p V_p = I_s V_s \quad \text{for 100\% efficiency}$$

A step-up transformer that increases the voltage by a factor of 100 must therefore lead to a decrease in the current by a factor of 100.

In practice the efficiency is always less than 100%. The efficiency of the transformer is given by

$$\text{efficiency} = \frac{\text{output power}}{\text{input power}} = \frac{I_s V_s}{I_p V_p}$$

Transformers are less than 100% efficient for the following reasons.

- Energy is transferred as heat in the windings. Power loss in the primary and secondary coils is due to the electrical resistance of the coils. The efficiency can be improved by reducing the resistance of the coils – by using thicker copper wire or even by using superconducting coils, as in some experimental transformers.
- Energy is transferred as heat in the ferromagnetic core. The changing flux in the core leads to '**eddy currents**' being induced in the core. The size of the eddy current is reduced by using a core that is laminated – sheets of iron are interleaved with sheets of an insulating material.
- The changing magnetic flux leads to fluctuating forces between the primary and secondary coils, and between the metal sheets in the laminated core. These varying forces lead to the buzzing that can sometimes be heard coming from transformers.
- Not all of the flux generated by the primary intercepts the secondary. This does not necessarily lead to energy losses, but if this leakage flux intercepts a conductor, there will be eddy currents. Flux linkage can be improved by using co-axial coils, where the secondary is inside the primary.
- There are also hysteresis losses in the core – as the core is magnetised and demagnetised, some energy is dissipated in heating it. These losses can be reduced by using new materials for the core which are easily magnetised and demagnetised.

Small transformers, such as those used for laptop computers or for charging mobile phones, are often less than 85% efficient. Larger transformers, such as those used in the National Grid, are likely to be better than 98% efficient.

The National Grid

Power stations that generate electricity are often a long distance from where the electricity is needed. A transmission system, the National Grid, is used to transmit the electricity. There are significant energy losses due to the resistance of the cables used. These energy losses are due to heating of the wires and are dependent on the current, I, that flows in the wires. The power loss due to resistance, R, is given by

$$P_{\text{loss}} = I^2R$$

In order to keep the power loss as low as possible, it is important to transmit the electrical power at low current. However, the power delivered from the power station is $P = IV$. To transmit high power, at low current, a very high potential is needed. Step-up transformers are therefore used at power stations to increase the voltage to a very high value, 400 kV, for transmission over long distances. Step-down transformers are used near to towns and industrial sites to reduce the potential to safer values.

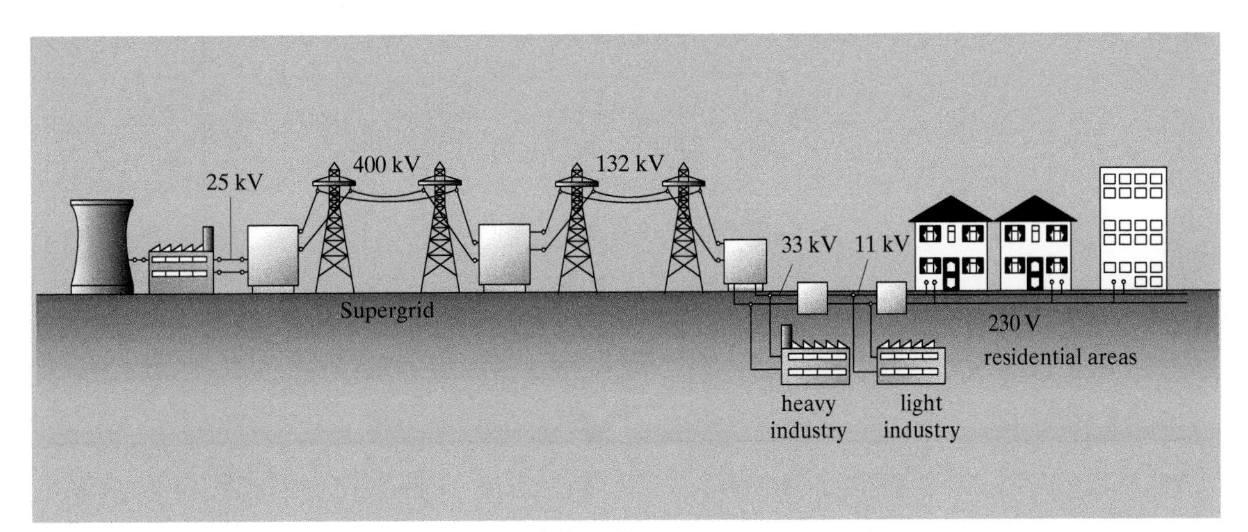

Fig 96
Electrical energy generated at 25 kV is stepped up for long-distance transmission via the Supergrid to 400 kV, then down to 132 kV for local distribution networks. This is stepped down further for use in factories, and down to the standard domestic voltage of 230 V for home use.

Example

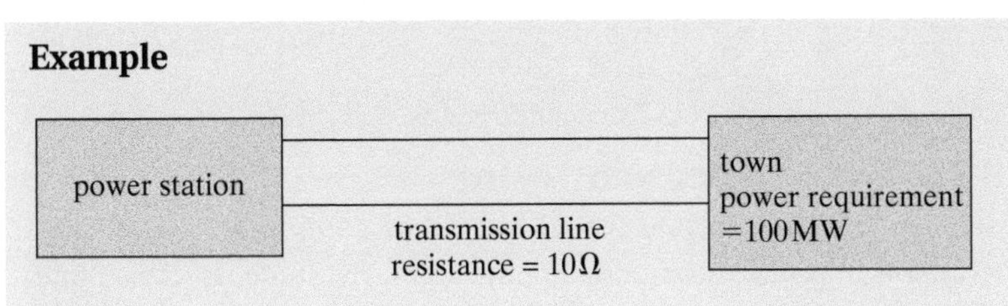

Fig 97

A power station is connected to a town by cables of resistance 10 Ω.

1 Calculate the power losses in the cables if:

(a) The power is supplied to the town at 20 kV.

(b) The power is supplied to the town at 400 kV.

2 Compare the efficiency of transmission at low voltage and at high voltage.

Answers

1 (a) The power required at the town is 100 MW. The current is found from $P = IV$, giving

$$I = \frac{P}{V} = \frac{100 \times 10^6}{20 \times 10^3} = 5000\text{ A}$$

Power losses in the cables = $I^2R = 5000^2 \times 10 = 250$ MW

(b) Similarly,

$$I = \frac{P}{V} = \frac{100 \times 10^6}{400 \times 10^3} = 250\text{ A}$$

Power losses in the cables = $I^2R = 250^2 \times 10 = 625$ kW

2 At low voltage the power station has to provide a total power of $100 + 250 = 350$ MW, so the efficiency is

$$\frac{\text{useful power}}{\text{total power supplied}} = \frac{100}{350} = 29\%$$

At high voltage the power station has to provide a total power of $100 + 0.625 = 100.625$ MW, so the efficiency is

$$\frac{\text{useful power}}{\text{total power supplied}} = \frac{100}{100.625} = 99.4\%$$

3.8 Nuclear physics

3.8.1 Radioactivity

3.8.1.1 Rutherford scattering

Working at Manchester University in 1909, Ernest Rutherford conducted experiments to probe the structure of the atom. His students, Geiger and Marsden, studied the scattering of alpha particles as they passed through a thin piece of gold foil (Fig 98). They detected the alpha particles using a scintillator, in this case a zinc sulfide screen. This emits a brief flash of light

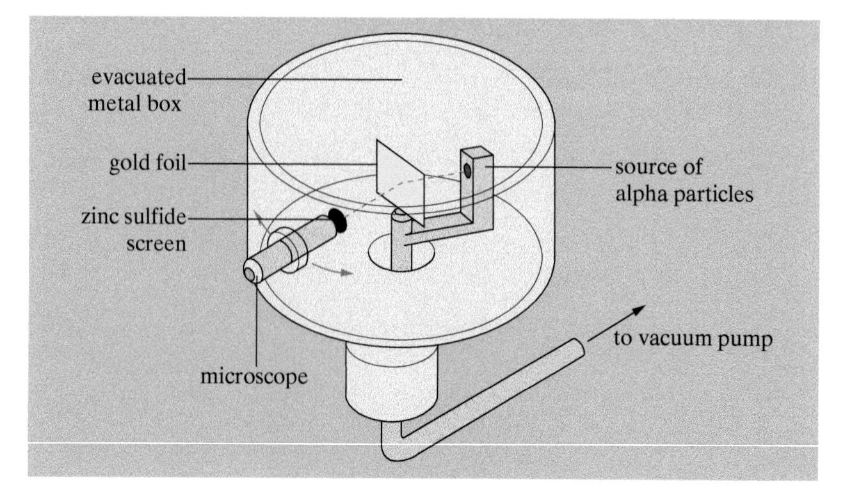

Fig 98
Rutherford scattering apparatus

when struck by an alpha particle. The detector was moved through a large angle. The number of alpha particles arriving at each angle was recorded. The significant results of the experiment were as follows:

- most alpha particles passed through the gold foil without deviation
- a significant number of particles were deflected by a small angle
- a very small number of alpha particles, around one in 8000, were deflected through a large angle, bounced back towards the source (see Fig 99).

Fig 99
Rutherford scattering

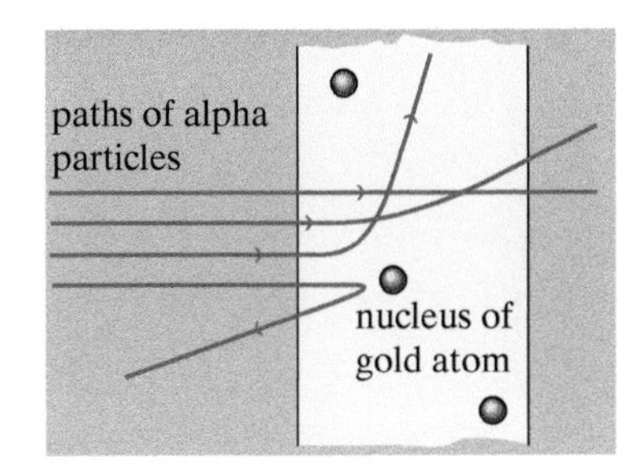

Rutherford drew the following conclusions:

- Atoms, of gold in this case, are almost all empty space.
- Almost all of the atom's mass must be concentrated in one small volume, which is now known as the nucleus.
- The nucleus is very dense and carries a net positive charge.
- The negative charge of the atom is carried by electrons, which have very low mass.
- The electrons orbit the nucleus at a distance that is much greater than the diameter of the nucleus itself.

In fact, the diameter of the atom is around 100 000 times greater than the diameter of the nucleus.

In 1909, the preferred model of the atom was the 'plum pudding' model put forward by J.J. Thomson and others. In the plum pudding model, the mass of the atom is uniformly distributed as a positive 'pudding' with electrons stuck in like plums. The results of the alpha scattering experiment could not be explained by the plum pudding model, and Rutherford's nuclear model became widely accepted.

Example

Write a short paragraph explaining why the observations made by Geiger and Marsden led to the abandonment of the plum pudding model.

Answer

The majority of alpha particles were not deflected at all, suggesting that atoms are almost all empty space. The small number of alpha particles that were deflected through a large angle suggests that the mass of an atom is concentrated in the nucleus, which must be positively charged (to repel the positively charged alpha particle).

3.8.1.2 α, β and γ radiation

Radioactivity was discovered by Henri Becquerel in Paris in 1897. He noticed that photographic film that had been stored in the dark near to uranium salts, turned black when developed, just as it would if it had been exposed to light. Becquerel concluded that the uranium salts were emitting invisible radiation, which could affect photographic film. Further experiments showed that the radiation could cause **ionisation** of molecules (in the air, for example). Marie and Pierre Curie identified other elements, such as radium, which also emit similar radiation.

Experiments by Becquerel, Rutherford and others identified three distinct types of ionising radiation, which are now referred to as **alpha** (α), **beta** (β) and **gamma** (γ) radiation.

Alpha radiation

Alpha radiation is the least penetrating of these radiations. It is made up of particles, which are positively charged and relatively massive. Rutherford showed that an alpha particle is identical to a helium nucleus. Each alpha particle is a tightly knit group of two neutrons and two protons, held together by the strong nuclear force. An alpha particle can be fired out from an unstable nucleus, usually a large nucleus like radium. The alpha particle is emitted with an energy of up to 10 MeV. The energy for this decay comes from a mass difference between the original, or **parent nucleus**, and the decay product, or **daughter nucleus**. Bismuth-212 is an alpha emitter that decays to thallium:

$$^{212}_{83}\text{Bi} \rightarrow ^{208}_{81}\text{Tl} + ^{4}_{2}\text{He}$$

Alpha particles have only a short range in air, typically 5 cm, and are stopped by a thin sheet of material such as paper or skin. Alpha radiation is intensely ionising, creating tens of thousands of ion pairs per cm in air.

Alpha radiation is not a major risk to health, provided that the emitter is *outside* the body, since its limited penetration means that all the energy is dissipated in the outer layer of skin. However, alpha emitters are very damaging when ingested, since all the energy is deposited in a small volume. Radon is an alpha-emitting radioactive gas that can accumulate in buildings, particularly in parts of the country with granite rocks, like Cornwall. Radon has been identified as increasing the risk of lung cancer.

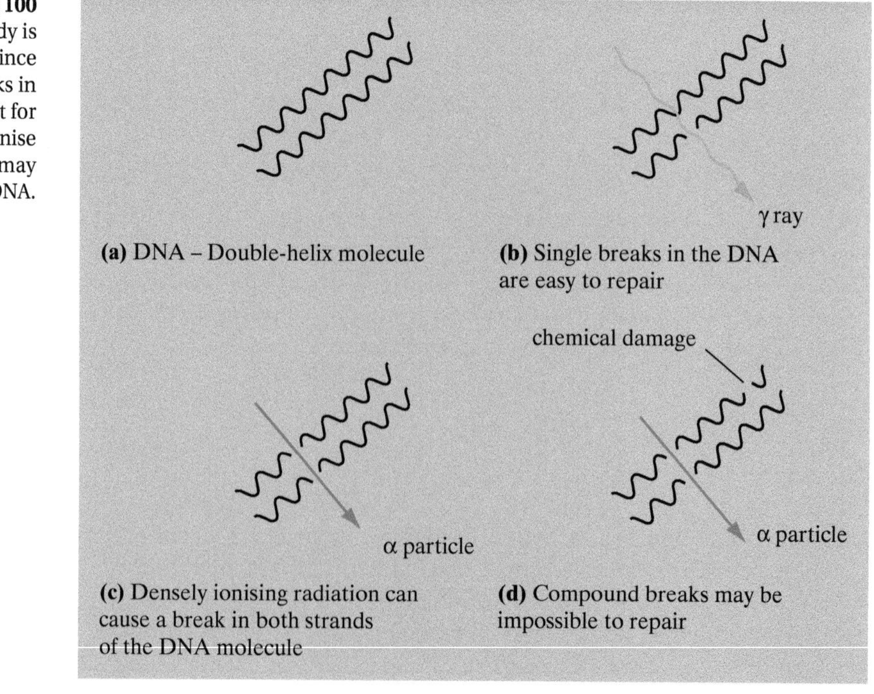

Fig 100
Alpha radiation inside the body is particularly damaging to DNA, since it is likely to cause multiple breaks in the double helix that are difficult for the cell to repair. It may also ionise other molecules in a cell, which may cause chemical damage to the DNA.

Alpha radiation from americium-241 is used in smoke detectors to cause an ionisation current in a small chamber. Smoke particles are larger and less mobile than air molecules, so when smoke particles enter the chamber they reduce this ionisation current. The drop in current causes an alarm to be sounded.

Beta radiation

Beta particles are very fast moving electrons that are emitted from the nucleus of some **radioisotopes**. This happens when a neutron decays into a proton and an electron. The proton remains inside the nucleus and the electron is emitted at close to the speed of light. An antineutrino, $\bar{\nu}_e$, is also emitted.

Strontium-90 is a beta emitter that is produced in nuclear reactors as one of the waste products of the nuclear fission of uranium. The equation describing its decay is

$$^{90}_{38}\text{Sr} \rightarrow {}^{90}_{39}\text{Y} + {}^{0}_{-1}\text{e} + \bar{\nu}_e$$

Essential Notes

Remember that isotopes are atoms of the same element, with the same number of protons in their nuclei but a different number of neutrons. **Radioisotopes** are radioactive elements whose nuclei have an unstable arrangement of nucleons (see page 111).

Beta particles have a longer range in air than alpha particles, typically 2 to 3 m. They are rather more penetrating and can travel through thin sheets of absorber such as plastic or paper. Beta radiation is less densely ionising than alpha radiation. Because beta particles are negatively charged, they are deflected by electric and magnetic fields.

Beta radiation presents a health hazard to humans when the source is outside the body, though the penetration is quite small. Most of the radiation damage is to the skin and surface tissues. A beta radiation source inside the body causes less local damage than an alpha source because its energy is dissipated over a larger volume. However, like all ionising radiation, a significant dose of beta radiation will damage DNA and lead to cell death, mutations or cancer.

Beta radiation is used to monitor the thickness of paper in a paper mill. If the paper is too thick, more beta radiation will be absorbed and the count rate will drop. An automatic adjustment can be made to the production process.

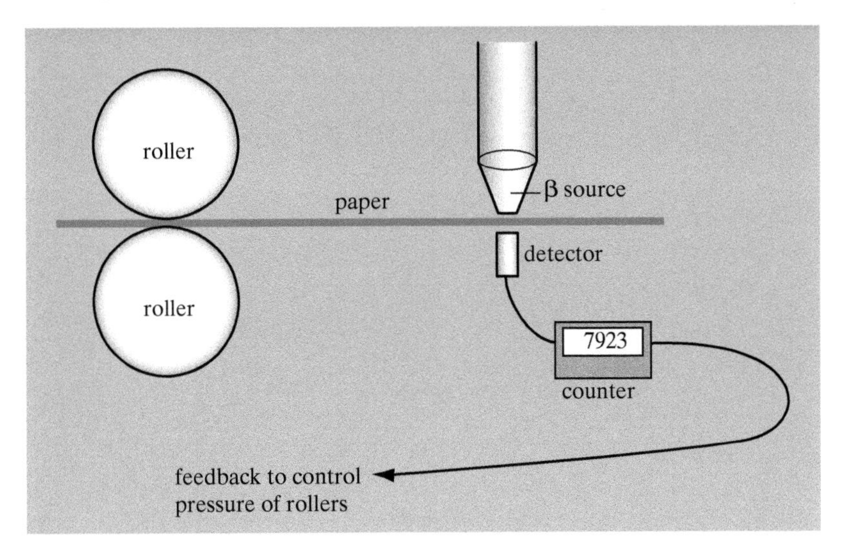

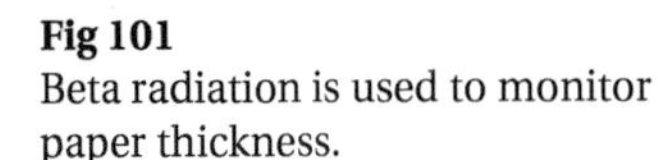

Fig 101
Beta radiation is used to monitor paper thickness.

Gamma radiation

Gamma radiation is very high-frequency electromagnetic radiation that is emitted from the nucleus of some radioisotopes. This emission does not change the nucleus to that of another isotope, but the nucleus does reduce its energy. An example of a gamma emitter is cobalt-60, which is often used in radiotherapy (see page 106). The cobalt-60 nucleus is in a higher energy state than normal and it decays to the **ground state**, emitting a gamma ray photon in the process:

$$^{60}_{27}\text{Co} \rightarrow ^{60}_{27}\text{Co} + \gamma$$

Gamma rays are not charged and so cannot be deflected by electric or magnetic fields. They interact with matter less strongly than alpha or beta particles and are much less ionising. As a result of this, gamma radiation is very penetrating and can easily pass through thin sheets of metal. In fact thick sheets of steel or lead are used to shield against gamma radiation. Even so, the shielding just reduces the intensity of the gamma radiation and does not completely absorb it. See Fig 102.

Gamma rays present a danger to human health even when the source is some distance from the body. A large dose of external gamma radiation could well have the same biological impact as a low dose of internal alpha particles. The best way of protecting yourself from gamma radiation is to keep well away from the source, since the intensity decreases with distance.

Gamma-emitting radioisotopes are often used as **tracers**. A small amount of radioactive gas can be introduced into a pipeline. If the gamma intensity above the ground is then measured, it is possible to find a leak. Gamma radiation is used in industrial radiography to produce 'shadow' pictures, in the same way that X-rays are used in medicine. A gamma-emitting source can be placed into a welded pipeline with photographic film around the outside. Any cracks in the welding will be shown as an overexposed line on the film.

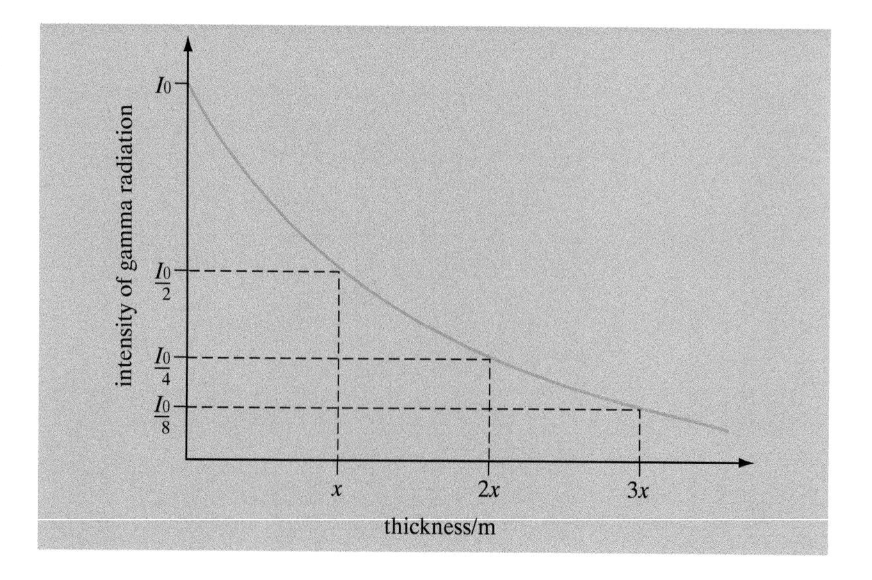

Fig 102
Exponential absorption of gamma radiation

For gamma rays of a given energy it takes a fixed thickness of shielding to reduce the intensity by half.

Identifying types of radiation

A Geiger counter is the main instrument used to detect ionising radiation in a school laboratory. Although alpha, beta and gamma radiations can be detected with a Geiger counter, it is most effective at detecting beta radiation. A Geiger counter is less effective at detecting alpha radiation, as most alpha rays do not penetrate the counter. Gamma radiation is less densely ionising than alpha or beta, and most gamma rays will pass through the Geiger counter without being detected.

It is possible to discriminate between the three types of radiation using the apparatus shown in Fig 103 by putting different absorbers in front of the Geiger counter. Alpha radiation will be stopped by a thin sheet of paper. A thin sheet of aluminium will stop beta radiation. Gamma radiation requires a significant thickness of lead to reduce its intensity.

Essential Notes

A Geiger counter detects the ionisation caused by radiation. Although a Geiger counter is quite effective at detecting beta radiation, it is much less efficient at counting alpha or gamma radiation, and so will significantly underestimate the activity of alpha or gamma emitters.

Essential Notes

Radioactive decay is a random event and the background count rate is likely to be low. To get an **accurate** measurement, you should count for a significant length of time.

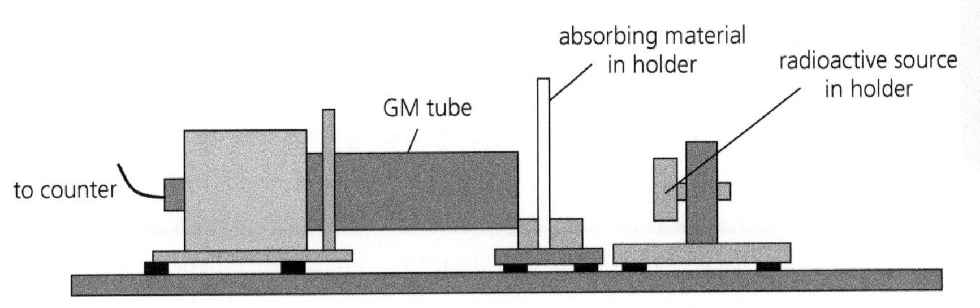

Fig 103
Measuring the absorption of ionising radiation

In all experiments involving radioactivity, it is important to take background radiation into account. Several readings of the background count should be taken, with the radioactive source(s) to be investigated placed well away from the detector. A mean value of background count can be found. This should be subtracted from all subsequent readings.

Working safely

The activities of the radioactive sources used in a school physics laboratory are quite low. Even so, it is important to work safely, as ionising radiation can present a significant health risk. The following recommendations should be followed.

- Secure storage. Radioactive sources should be kept in a suitable, lead-lined container and stored in a locked cupboard well away from people.
- Work quickly to reduce your exposure time.
- Keep your distance. Radiation dose decreases as you get further from the source.
- Do not touch. Avoid contamination. Do not handle the source directly. Use long tongs.

The inverse square law

The **intensity**, I, of gamma radiation at any point is the power that flows through an area of one square metre. Intensity is measured in watts per square metre, W m^{-2}. The gamma radiation from a small source can be considered to be the same in all directions (isotropic) and therefore

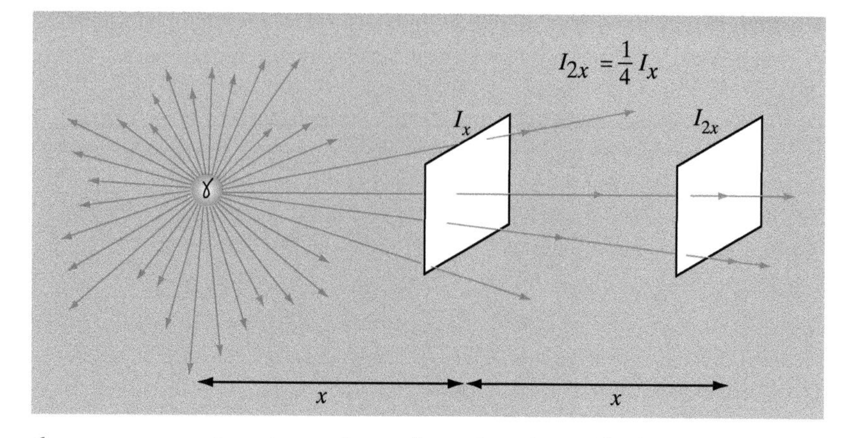

Fig 104
Inverse square law for gamma radiation

Any radiation that emanates from a point source, such as light from a small bulb, will follow the inverse square law.

Essential Notes

The intensity of alpha and beta radiation does not follow the inverse square law, because these particles are absorbed and deflected by molecules of air to a much greater extent than gamma radiation. In a vacuum the law would apply equally to all three of these forms of radiation.

the energy spreads out over the surface of a sphere. The intensity at a distance x away from a source of constant intensity is given by the inverse square law (see Fig 104):

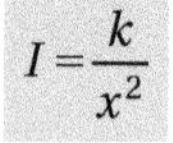

$$I = \frac{k}{x^2}$$

where k is a constant.

This has important implications for safe handling of a gamma source. The inverse square law means that doubling the distance from the source decreases the radiation dose to one-quarter. It is important to use long-handled tongs to manipulate the source. Gamma sources should be stored well away from people.

Example

At a distance of 20 cm from a gamma source the intensity is 1.6 μW m^{-2}. At what distance will the intensity drop to 0.1 μW m^{-2}?

Answer

Use the equation

$$I = \frac{k}{x^2}$$

At a distance of 0.20 m

$$1.6 \times 10^{-6} = \frac{k}{0.2^2}$$

Apply the equation again at a distance x

$$0.1 \times 10^{-6} = \frac{k}{x^2}$$

If we divide these equations,

$$\frac{1.6}{0.1} = \frac{x^2}{0.22}$$

$$x^2 = 0.64$$

$$x = 0.80 \text{ m}$$

Demonstrating the inverse square law

The inverse square law for gamma radiation can be demonstrated using a Geiger counter to measure the count rate. The count rate has to be corrected by subtracting the background count from each reading.

In a school laboratory a Geiger-Müller tube, known as a GM tube, and an electronic counter, known as a digicounter, are used to measure count rate.

The **corrected count rate**, C, is recorded at a number of distances, x, from the gamma source. There is a systematic error in the measurement of x, because it is difficult to measure exactly where the gamma rays are emitted from and where they are detected. This uncertainty adds an average distance d to the distance (see Fig 105). If the inverse square law is correct then

$$C \propto \frac{1}{(x+d)^2}$$

So, introducing a constant k,

$$x = kC^{-1/2} - d$$

Compare this to the equation for a straight line, $y = mx + c$. A graph of distance, x, on the y-axis against $C^{-1/2}$ on the x-axis should yield a straight line with a gradient of k and an intercept of $-d$.

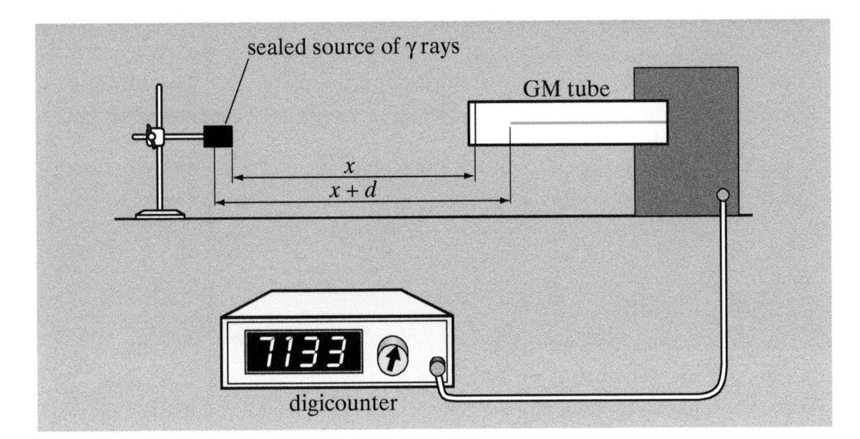

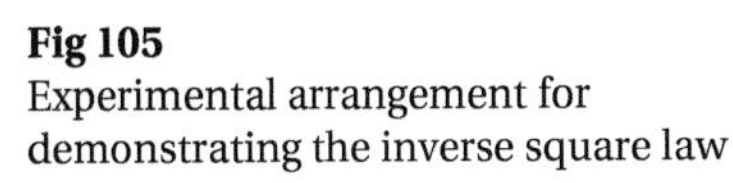

Fig 105
Experimental arrangement for demonstrating the inverse square law

Summary of properties of α, β and γ radiation

Table 6

Radiation	Nature	Penetrating power	Range in air	Ionising effect	Behaviour in electric and magnetic fields
Alpha	Two protons and two neutrons (helium nucleus)	Easily stopped, e.g. by a sheet of paper or the outer layer of (dead) skin cells	A few cm	Intensely ionising: an alpha particle will cause about 10^4 to 10^5 ion pairs per cm in air	Positively charged, so deflected by electric and magnetic fields; but relatively massive, so deflected less than beta particles
Beta	An electron	Stopped by thin (a few mm)* metal sheet	Several metres	Less intensely ionising than alpha: a beta particle will cause about 1000 ion pairs per cm*	Negatively charged, so deflected in opposite direction to alpha; deflected more than alpha as the mass is much less
Gamma	High-frequency electro-magnetic radiation	Reduced in intensity by $\frac{1}{2}$ by about 5 cm of concrete or 1 cm of lead*	10 to hundreds of metres*	Weakly ionising: about 10 ion pairs per cm*	Not charged, so undeviated by a magnetic or an electric field

*The actual value depends on the initial energy with which the radiation is emitted.

Background radiation

We live in a radioactive world. Every day we are exposed to nuclear radiation from the air that we breathe and the rocks that we walk on. We are also exposed to a small radiation dose due to medical and industrial procedures, though the largest fraction (around 87%) of the dose is from natural sources. See Fig 106.

The major sources of background radiation dose are:

- **Air.** This dose is mainly from radon and thoron, radioactive gases that are part of long-lived decay series.

- **Rocks and buildings.** Some rocks, particularly granite, contain uranium-238 or thorium-232. These two isotopes have long half-lives and decay to other radioactive products. The radiation dose from this source depends on where you live.
- **Cosmic rays.** The Earth is continually being bombarded with high-energy particles and gamma rays from space, mainly from the Sun but also from other sources outside the Solar System. Fortunately the atmosphere screens us from the worst effects of this radiation.
- **Food and drink.** Radioactive isotopes dissolved in water are taken up by plants, and then animals. Some foods are naturally more radioactive than others because they tend to concentrate radioactive isotopes. People are radioactive mainly due to the isotope potassium-40, which is concentrated in muscle.
- **Medical** procedures also contribute to our radiation dose. Most of this is due to diagnostic X-rays, but nuclear medicine techniques that use radioisotopes are becoming more common (see page 111).
- **Miscellaneous.** Artificial background radiation also comes from industrial techniques, mining wastes, nuclear power, and even from the fall-out from nuclear weapons tests that were carried out in the 1950s and 1960s.

Essential Notes

Most naturally occurring radioisotopes are part of a decay series – a chain of radioactive decays in which a radioactive element decays to another radioactive element, and so on until a stable isotope is attained. The intermediate radioisotopes may have half-lives from millions of years to just minutes.

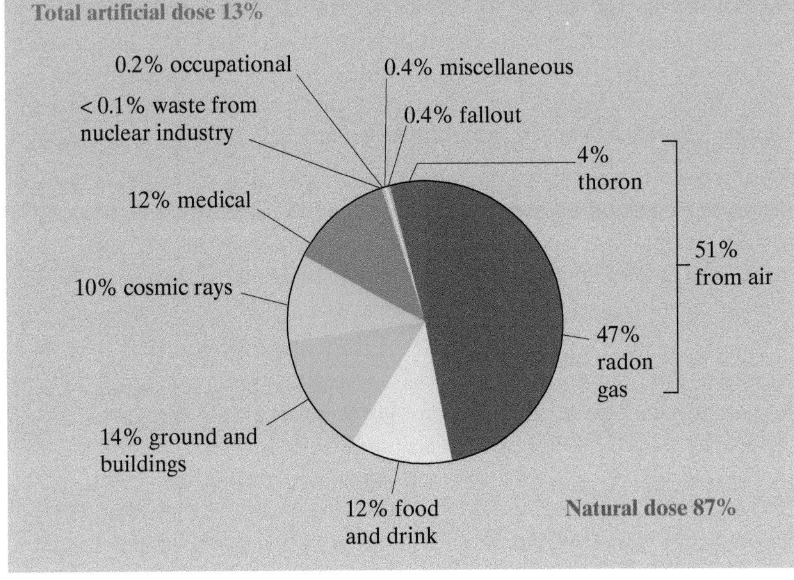

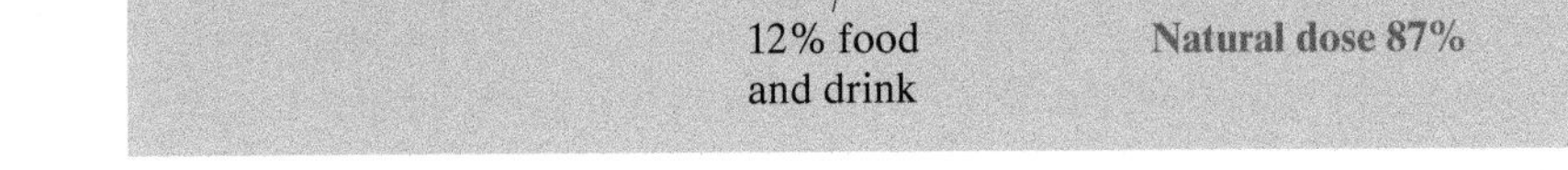

Fig 106
Sources of background radiation dose

The **background radiation count rate** has to be taken into account when measurements of radioactivity are undertaken in the laboratory.

Radioisotopes in medicine

Radioisotopes are used in the diagnosis and treatment of disease. For example, PET (positron emission tomography) scanning uses a tracer, a compound labelled with a very small amount of a radioactive isotope. Parkinson's disease can be diagnosed using an injection into a patient's bloodstream of 18-fluoradopa, a radioactively labelled compound of dopamine. 18-fluoradopa emits positrons, which annihilate electrons, emitting gamma rays in the process. The technique enables clinicians to assess how many dopamine neurons (nerve cells) a patient has lost due to disease. The radiation dose to the patient is minimal.

Iodine-131, a beta emitter, is used as a tracer to diagnose and treat hyperthyroidism, a disease in which a person's thyroid gland becomes over-active. Most of the iodine taken in by the body is concentrated within the thyroid gland. If radioactive iodine-131 is injected into the patient, the radiation produced by the iodine that concentrates in the thyroid gland is monitored to assess thyroid function. A larger dose can be given to kill cells in the thyroid gland, to reduce the thyroid function and cure the over-activity.

Kidney function can also be assessed by using a tracer. A gamma emitter, such as technetium-99, is injected into a patient's bloodstream and a gamma camera is used to monitor the absorption and excretion of the gamma emitter by the kidneys.

Gamma-emitting isotopes are usually used in radiotherapy to kill cancerous cells, since gamma radiation can penetrate deep into the body. The radioisotope is kept in special machinery that can direct the radiation to particular parts of the body, as well as shield much of the rest of a patient and the operator from anything more than a minimal dose. Radiotherapists design treatment plans to maximise the radiation dose to cancer cells, whilst keeping the dose to healthy cells as low as possible.

Alpha-emitting radioisotopes are less commonly used in medicine, though the short range and high ionising power of alpha radiation can be useful in killing cancer cells, if the radioisotope can be delivered to an exact location. Trials have been conducted using radium-223 in an attempt to treat bone cancer, since active bone cells take up radium.

3.8.1.3 Radioactive decay

Radioactive decay is a *random* event. For a given sample of material, there is no way of predicting with any certainty which nucleus will decay, or when it will decay. All we can do is to give the **probability** that a nucleus will decay in a given time. Some radioisotopes decay relatively quickly, whereas some take thousands of years. For every radioisotope the probability that a nucleus will decay in a certain time is given by the **decay constant**, λ. In a large number of radioactive nuclei, N, the number of atoms that decay in a certain time, Δt, is ΔN. The probability that any particular nucleus will decay in time Δt is

$$\frac{-\Delta N/N}{\Delta t} = \lambda$$

This equation is usually written as

$$\frac{\Delta N}{\Delta t} = -\lambda N$$

or

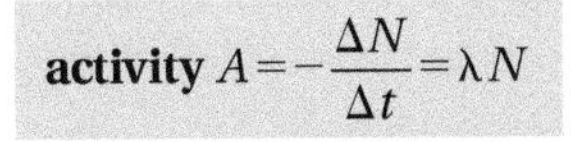

$$\textbf{activity } A = -\frac{\Delta N}{\Delta t} = \lambda N$$

This equation says that the activity of a radioactive source, $\Delta N/\Delta t$, is proportional to the number of active nuclei present. The decay constant λ is the constant of proportionality, which is different for each radioisotope.

Radioactivity is also *spontaneous*. The rate of decay is not influenced by external factors, such as temperature or pressure.

Activity

The **activity** of a radioactive source measures the number of decays that occur, on average, every second. Activity is measured in **becquerel**, Bq. One becquerel is one disintegration per second. The activity of a source is measured with a Geiger counter.

Exponential decay

Consider a sample of a radioisotope. As time goes by the number of radioactive nuclei in the sample of material will reduce. At first there are a large number of nuclei and so the decay rate is also high. As the number of nuclei reduces, the decay rate also reduces. This kind of behaviour is known as **exponential decay**.

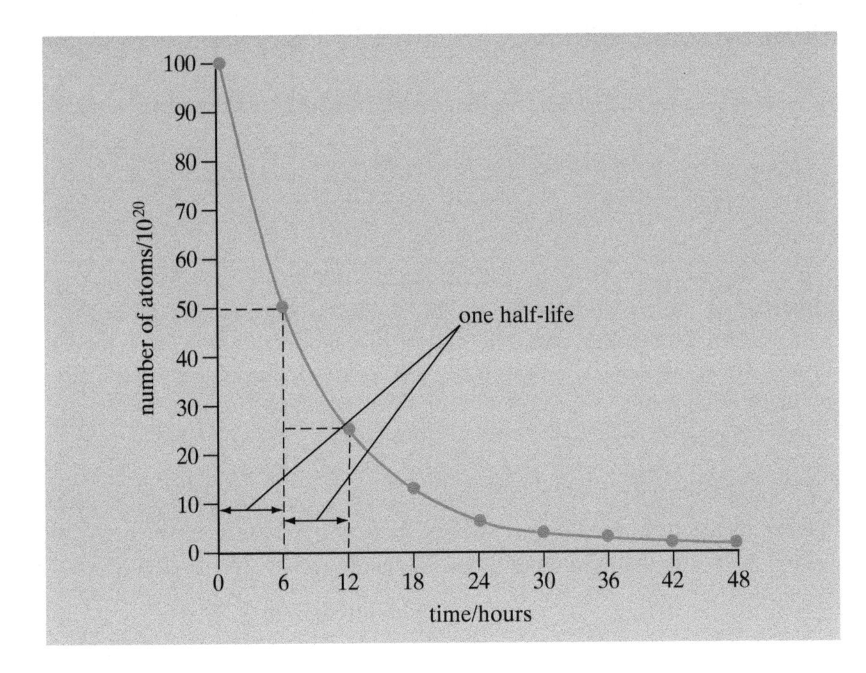

Notes

The negative sign in these equations indicates that the change in the number of atoms, ΔN, is a *decrease*.

Essential Notes

Dice can be used to model radioactive decay. Each die represents an atom of a radioactive isotope. Suppose we throw a large number of dice, e.g. 6 000. We cannot predict which die will show 6, but we can predict that roughly 1 in 6, around 1 000 in this case, will show 6. Similarly we cannot tell which atoms will decay, but we can predict approximately how many will decay in a certain time.

Essential Notes

You have met another example of exponential decay in 3.7.4.4: the charge vs time graph for a discharging capacitor follows the same exponential law.

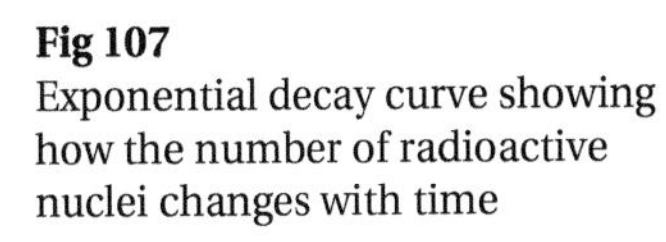

Fig 107
Exponential decay curve showing how the number of radioactive nuclei changes with time

Notes

It is usual to use t in seconds and λ in units of s^{-1}. However sometimes it is more convenient to use time in years. λ must then have units of $(years)^{-1}$.

The equation that describes the graph in Fig 107 is

$$N = N_0 e^{-\lambda t}$$

where N_0 is the initial number of atoms present, i.e. the number at time $t = 0$ s, λ is the decay constant, and t is the time.

The activity of a source, measured in becquerels, is also proportional to the number of active nuclei present: $A = \lambda N$. The activity of a source therefore follows the same exponential decay law:

$$A = A_0 e^{-\lambda t}$$

Example

Radioactive iodine-131 has a decay constant of $9.9 \times 10^{-7}\ s^{-1}$.
A sample of iodine-131 has an initial mass of 0.10 g.

(a) Calculate the initial activity of the source.

(b) Calculate the activity after 30 hours.

Answer

(a) The activity, A or $\Delta N/\Delta t$, is given by $-\Delta N/\Delta t = \lambda N$.

The number of atoms present in a mole of iodine is 6.02×10^{23} (this is the Avogadro constant, see page 41).

A mole of iodine-131 would have a mass of 131 g. So 0.10 g represents $0.10/131 = 7.63 \times 10^{-4}$ of a mole, which is $7.63 \times 10^{-4} \times 6.02 \times 10^{23} = 4.60 \times 10^{20}$ atoms.

So the activity is

$$\lambda N = 9.9 \times 10^{-7} \times 4.60 \times 10^{20} = 4.55 \times 10^{14}\ \text{Bq}$$

(b) The activity after 30 hours follows the rule $A = A_0 e^{-\lambda t}$.

$t = 30\,h = 30 \times 3600 = 1.08 \times 10^5\,s$

So $A = 4.55 \times 10^{14} \times e^{-(9.9 \times 10^{-7} \times 1.08 \times 10^5)}$

$= 4.09 \times 10^{14}$ Bq

Essential Notes

1Bq is equivalent to $1s^{-1}$

Half-life

Every radioactive isotope has its own **half-life**. This is the time taken for the number of active nuclei in a sample of radioactive material to drop to half of the original value.

Definition

The half-life is the time it takes for the number of active nuclei in a sample to drop to half of its original value.

The half-life is also the time it takes for the activity to drop to half of its original value. After two half-lives the activity will drop to $\frac{1}{4}$ of its original value and after three half lives the activity will only be $\frac{1}{8}$ of the original activity.

The half-life is linked to the decay constant, λ. An isotope with a small value of λ has nuclei with a low probability of decay, so the isotope has a long half-life.

The exact relationship can be deduced from the equation $N = N_0 e^{-\lambda t}$. After one half-life N will equal $N_0/2$:

$$\frac{N_0}{2} = N_0 e^{-\lambda t} \quad \text{or} \quad \frac{1}{2} = e^{-\lambda t}$$

Taking logarithms of both sides of this equation gives

$$\ln\left(\frac{1}{2}\right) = -\lambda t$$

Since $\ln\left(\frac{1}{2}\right) = -\ln 2$

$$\text{half-life } T_{\frac{1}{2}} = \frac{\ln 2}{\lambda}$$

It is possible to plot a logarithmic–linear (see page 127) graph to find the half-life of an isotope. Suppose that we measure the activity, $A = -\Delta N/\Delta t$, at several different times, t.

Since $A = A_0 e^{-\lambda t}$, taking logarithms of both sides gives

$$\ln A = \ln (A_0 e^{-\lambda t})$$

This becomes

$$\ln A = \ln A_0 + \ln (e^{-\lambda t})$$

$$\ln A = \ln A_0 - \lambda t$$

Compare this with the equation of a straight line, $y = c + mx$. If we plot ln A on the y-axis and t on the x-axis, we will get a straight line with a gradient of $-\lambda$. Since λ is the decay constant, we can then calculate the half-life from the gradient by using the relationship

$$T_{\frac{1}{2}} = \frac{\ln 2}{\lambda}$$

Essential Notes

We have used the fact that the logarithm of a product is equal to the sum of the logs of the individual numbers,

$\log(A \times B) = \log A + \log B$

Storage of radioactive waste

Some of the isotopes created as fission products in nuclear reactors are highly radioactive. These isotopes have a high value of λ and a short half-life. The spent fuel rods are stored under water in large ponds for several months until the activity from these isotopes has reduced (see page 124). Some of the other products, like plutonium-239, are much longer lived. Plutonium-239 has a half-life of over 20 000 years and presents long-term storage problems.

Example

Strontium-90 is one of the radioactive isotopes that are produced as waste products in fission reactors. The activity of a small sample of strontium-90 was measured over a period of 10 years and the following results were recorded.

Time/years	Activity/MBq
0	200.0
1	195.1
2	190.3
3	185.7
4	181.1
5	176.7
6	172.4
7	168.2
8	164.1
9	160.1
10	156.1

Plot a suitable graph to calculate the decay constant, and hence the half-life, of strontium-90.

Answer

Since $A = A_0\,e^{-\lambda t}$, taking logarithms gives $\ln A = \ln A_0 - \lambda t$. This is in the form $y = mx + c$, so we can plot $\ln A$ against the time, t. The gradient will give us the decay constant.

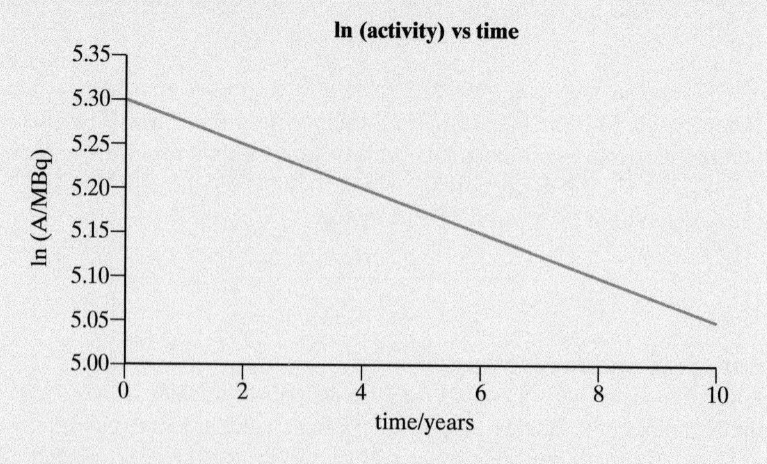

Fig 108
Using a logarithmic–linear graph to find half-life

The gradient is equal to -0.025 years^{-1}, so the half-life is $(\ln 2)/0.025 = 27.7$ years.

Radiocarbon dating

The radioactivity of a carbon isotope, carbon-14, is used to date artefacts that are made from organic materials. Carbon-14 is formed in the upper atmosphere by the action of cosmic rays on nitrogen. When nitrogen absorbs a neutron it decays to carbon-14:

$$^{14}_{7}\mathrm{N} + ^{1}_{0}\mathrm{n} \rightarrow ^{14}_{6}\mathrm{C} + ^{1}_{1}\mathrm{H}$$

The carbon-14 forms part of the atmospheric carbon dioxide and it becomes assimilated into plants through photosynthesis. Every living plant contains a small, known, percentage of radioactive carbon. During a plant's life the level of radioactive carbon stays more or less constant, the plant takes in new carbon-14 to replace that which has decayed. However, as soon as the plant dies the level of radioactivity from carbon-14 begins to drop. Since we know that carbon-14 has a half-life of 5730 years, it is possible to estimate the age of the artefact from the activity.

Essential Notes

The electrostatic force has a much greater range than the strong nuclear force, which is limited to around 10^{-15} m. Two protons on the opposite side of a large nucleus still repel each other due to the electrostatic force, but are out of range of the strong nuclear force.

3.8.1.4 Nuclear instability

Stable isotopes with low atomic numbers tend to have equal numbers of neutrons and protons in their nuclei. The nuclei are stable because the attraction of the strong nuclear force, which acts between all nucleons, is strong enough to balance the repulsive electrostatic force. This is not the case for isotopes with higher atomic numbers. Because these have more protons there is a greater electrostatic force pushing the nucleus apart. More neutrons are needed to 'glue' these nuclei together. Stable isotopes with large nuclei therefore have more neutrons than protons.

The stable nuclei are shown in Fig 109. Nuclei that lie either above or below this curve tend to be unstable.

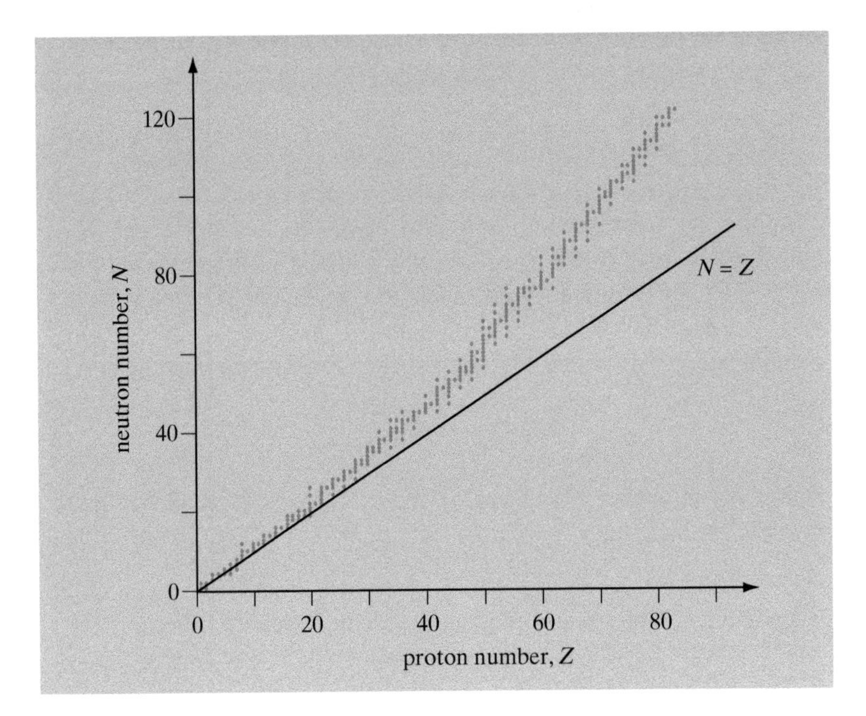

Fig 109
Graph of neutron number, N, against proton number, Z

Vertical lines represent isotopes. Isotopes which are further from the main curve tend to be more unstable.

Possible decay modes of unstable nuclei

There are a number of ways in which unstable nuclei can decay.

Alpha (α) decay

Alpha emitters are proton-rich heavy isotopes ($Z > 60$). An alpha particle is a very stable combination of two protons and two neutrons, identical to a helium nucleus, that can form within a large nucleus. After many attempts, the alpha particle may escape from the nucleus.

Alpha decay reduces the proton number, Z, of the parent nucleus by 2, and reduces the nucleon number, A, by 4. The general equation for alpha decay is

$$^{A}_{Z}\mathrm{X} \rightarrow {}^{A-4}_{Z-2}\mathrm{Y} + {}^{4}_{2}\mathrm{He}$$

where X is the parent (alpha-emitting) isotope and Y is the daughter. A gamma ray is often also emitted (see page 99).

Essential Notes

This is a random process, though the probability of emission is greater if the alpha particle has more energy. In fact there is a link between alpha particle energy and the half-life of the isotope; the greater the alpha's energy, the shorter the half-life.

Beta minus (β^-) decay

Beta decay occurs when a neutron is transformed into a proton, with the emission of a high-speed electron and an antineutrino from the nucleus.

Beta decay leaves the nucleon number unchanged, since a neutron has been transformed into a proton. The proton number goes up by 1. The general equation for beta-minus decay is

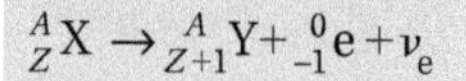

$$^{A}_{Z}\mathrm{X} \rightarrow {}^{A}_{Z+1}\mathrm{Y} + {}^{0}_{-1}\mathrm{e} + \nu_e$$

Beta-minus emitters are neutron-rich and lie above the curve of stable nuclei in Fig 10. A free neutron will decay into a proton and an electron (beta-minus) with a half-life of around 11 minutes. This decay is energetically possible because the mass of a neutron is slightly greater than the combined mass of the proton and electron.

Essential Notes

The ν_e is an antineutrino. Its existence makes sure that the laws of energy and momentum conservation are observed (see AS/A-Level Year 1, Section 3.2.1.2).

Beta plus (β^+) decay

Beta plus decay occurs when a proton is transformed into a neutron, emitting a high-speed positron and a neutrino from the nucleus. The positron is the antimatter version of the electron. It has exactly the same mass as the electron. The positron's charge is $+1.6 \times 10^{-19}$ C, equal in size but opposite in sign to the charge of the electron.

Beta-plus decay leaves the nucleon number unchanged, but the proton number goes down by one. The general equation for beta-plus decay is

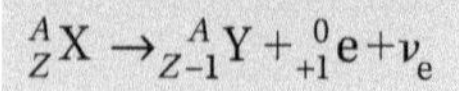

$$^{A}_{Z}\mathrm{X} \rightarrow {}^{A}_{Z-1}\mathrm{Y} + {}^{0}_{+1}\mathrm{e} + \nu_e$$

Beta-plus emitters are situated below the N vs Z curve for stable nuclei (Fig 110).

Essential Notes

The decay of a proton into a neutron and a positron only happens inside certain unstable nuclei. A free proton is a very stable particle.

Electron capture

It is possible for the nucleus of an atom to capture one of the atom's orbiting electrons. When this happens a proton in the nucleus absorbs the electron and becomes a neutron. This is how the isotope beryllium-4 decays:

$$^{7}_{4}\text{Be} + ^{0}_{-1}\text{e} \rightarrow ^{7}_{3}\text{Li} + \nu_e$$

Example

Two of the isotopes of thorium, $^{232}{}_{90}\text{Th}$ and $^{228}{}_{90}\text{Th}$, are found in the same sample of rock, both isotopes are unstable and decay by emitting an alpha particle. It has been suggested that thorium-232 decays into Th-228 via two other isotopes. Explain how this could happen and construct the decay chain, showing all the intermediate stages:

Answer

$$^{232}_{90}\text{Th} \rightarrow ^{228}_{88}\text{Ra} + \alpha$$

$$^{228}_{88}\text{Ra} \rightarrow ^{228}_{89}\text{Ac} + \beta + \bar{\nu}$$

$$^{228}_{88}\text{Ac} \rightarrow ^{228}_{90}\text{Th} + \beta + \bar{\nu}$$

The thorium-232 emits an alpha particle and becomes radium. Radium is a beta emitter and decays into actinium, which is also a beta emitter and this decays back to thorium.

Existence of nuclear excited states

Alpha emission is often accompanied by gamma-ray emission. After the alpha emission the daughter nucleus is left in an excited state. At some point this excited nucleus will decay by the emission of a gamma ray. The energy available for the decay is therefore shared between the alpha particle and the gamma ray.

The gamma rays emitted following alpha emission have a line spectrum that reflects the energy levels in the daughter nucleus.

Essential Notes

The daughter nucleus also has some kinetic energy due to its recoil after the decay. However, the mass of the daughter tends to be much larger than the alpha particle. The daughter moves at a much lower velocity than the alpha and so it carries much less kinetic energy.

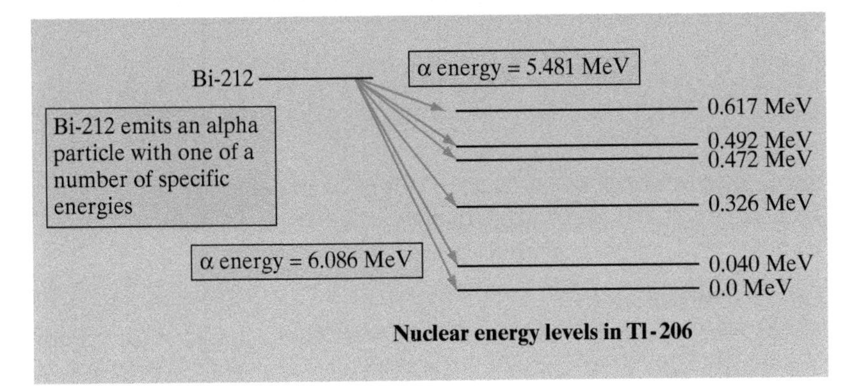

Fig 111
Bismuth-212 decays to thallium-206 by alpha emission. The decay can leave thallium in an excited state. The thallium later emits a gamma ray to return to the ground state.

Sometimes the gamma emission can be delayed by a significant time. The isotope is then marked with an 'm' to indicate that it is a **metastable state.** Technetium-99m is such an isotope. This is created by the decay of molybdenum-99, and has a half-life of 6 hours. The half-life of technetium-99m, and the fact that it is a gamma emitter, makes it ideal for use in hospitals as a tracer and for imaging.

Essential Notes

A radioisotope used for medical diagnosis needs to have a half-life long enough to enable the investigation to be carried out, but not too long or the activity will be low and the patient will remain radioactive for too long. Gamma emitters are used because the radiation is not absorbed by the body.

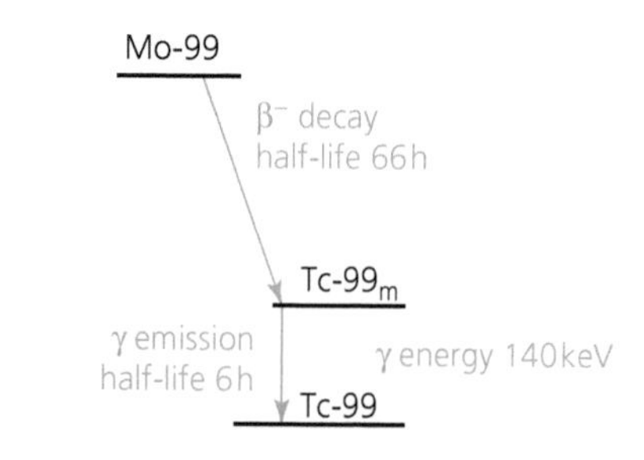

Fig 112
Simplified energy level diagram showing the formation of technetium-99m

3.8.1.5 Nuclear radius

Closest approach of an alpha particle

In an alpha scattering experiment such as Rutherford's (see page 96), when an alpha particle approaches a gold nucleus on a direct collision course it slows down and stops for an instant before rebounding. The alpha particle's kinetic energy is transferred to potential energy as the positively charged alpha particle does work against the electrostatic repulsion of the positively charged gold nucleus.

At the distance of closest approach, all the kinetic energy has been transferred to potential energy:

$$E_k = E_p$$

For a spherical charge, Q, the potential energy is

$$E_p = qV = \frac{Qq}{4\pi\varepsilon_0 r}$$

where V is the electric potential a distance r from the nucleus of charge Q (see 3.7.3.3).

For a 5 MeV alpha particle approaching a gold nucleus of atomic number 79, this becomes

$$(5\times10^6\times1.6\times10^{-19})\text{J} = \frac{(79\times1.6\times10^{-19})\text{C}\times(2\times1.6\times10^{-19})\text{C}}{4\pi\times8.85\times10^{-12}\text{Fm}^{-1}\times r}$$

giving a value for r of 4.55×10^{-14}m. This gives an upper limit for the radius of the gold nucleus. Modern measurements give a value of 6.5×10^{-15} m, or 6.5 fm, **femtometres**.

Electron diffraction

All particles have a wave-like nature. The wavelength, λ, of an electron can be calculated from the momentum, p, of the electron using de Broglie's relation:

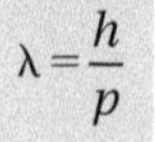

$$\lambda = \frac{h}{p}$$

where h is the Planck constant.

High-energy electrons can have a wavelength that is small enough to diffract around the nucleus. We can estimate the nuclear radius from these diffraction patterns.

Essential Notes

Wave–particle duality was discussed in AS/A-Level Year 1, Section 3.2.

Electrons can be accelerated by allowing them to pass through a potential difference, V. The larger the potential difference, the more energy the electron will gain and the smaller its wavelength will be.

The energy gained by the electron is $E = eV$, where $e = 1.6 \times 10^{-19}$ C is the charge on the electron. This energy will be equal to the final kinetic energy of the electron, $E_k = \frac{1}{2}mv^2$:

$$eV = \frac{1}{2}mv^2$$

This gives $m^2v^2 = 2\ meV$

So
$$p = mv = \sqrt{2meV}$$

The **de Broglie wavelength** is therefore

$$\lambda = \frac{h}{\sqrt{2meV}}$$

A potential difference of 100 V will give rise to electrons with a wavelength of 1.23×10^{-10} m.

Essential Notes

Compare this electron wavelength to the shortest wavelength of visible light, which is 400 nm or 4×10^{-7} m. The wavelength of the electron is more than 1000 times shorter.

By using higher voltages we can diffract an electron beam around the nucleus of an atom. The electrons are fired at a thin slice of material to produce a diffraction pattern.

The angle at which the first minimum of such a diffraction pattern appears (see Fig 113) is given by

$$\sin\theta = \frac{0.61\,\lambda}{R}$$

where R is the radius of the obstacle.

Electrons with an energy of 125 GeV have a de Broglie wavelength of 3.47×10^{-15} m, or 3.47 fm. An angle θ of 38° for oxygen nuclei gives a value for the radius of an oxygen nucleus of:

$$R = \frac{0.61 \times 3.46 \times 10^{-15}\ \text{m}}{\sin 38^\circ}$$
$$= 3.4 \times 10^{-15}\ \text{m}$$

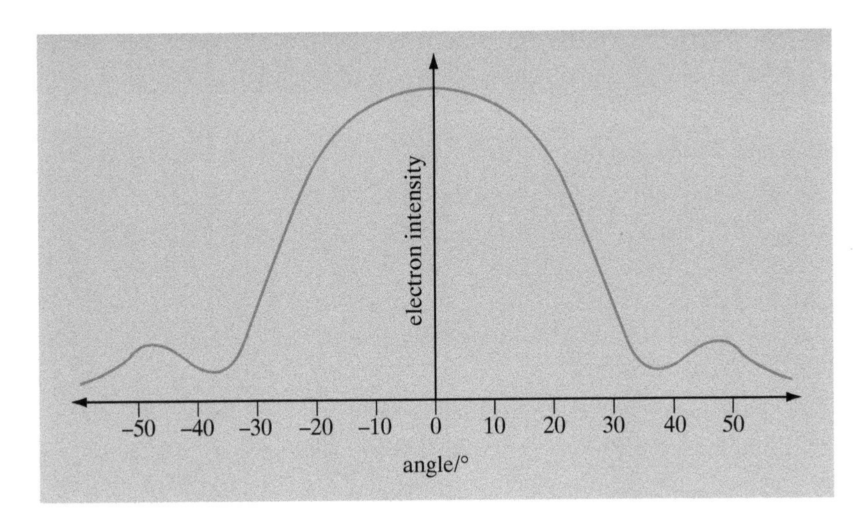

Fig 113
Electron diffraction around a nucleus

The first diffraction minimum due to an oxygen nucleus is at about 38°.

Diffraction experiments using different elements have shown that there is a link between the nucleon number, *A*, of a nucleus and its radius, *R*:

$$R = r_0 A^{1/3}$$

where r_0 is a constant representing the radius of a single nucleon.

Example

The following data are taken from an electron diffraction experiment.

Nucleon number, A	Radius R/fm
12	3.04
16	3.41
28	3.92
40	4.54
51	4.63

Plot a graph to verify that the nuclear radius, *R*, depends on the atomic number, *A*, according to this equation $R = r_0 A^{1/3}$.

Use your graph to find the value of r_0.

Answer

Because we are testing the relationship $R = r_0 A^{1/3}$, we need to plot the radius, *R*, against $A^{1/3}$ to get a straight line. The gradient will be the value of r_0.

Fig 114

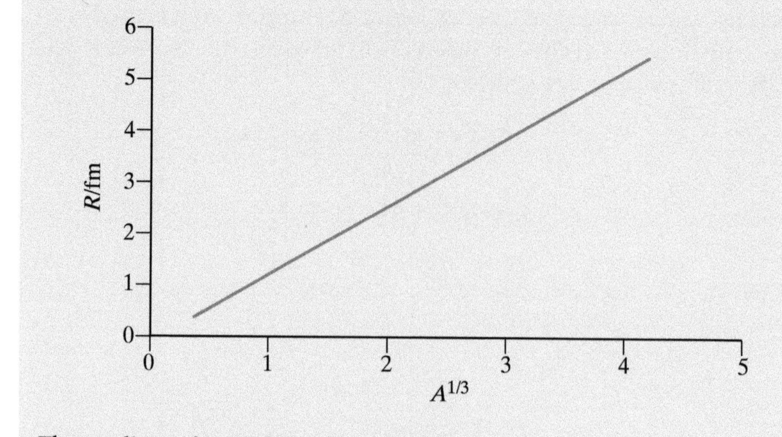

The gradient of 1.003 fm is the value of r_0.

Essential Notes

Density, $\rho = \frac{\text{mass}}{\text{volume}}$

Nuclear density

We can use the expression for the nuclear radius to investigate a value for the density of a nucleus.

The mass of a nucleus is approximately equal to $A \times m_n$, where *A* is the nucleon number and m_n is the mass of one nucleon. (There is a small difference between the mass of a proton and the mass of a neutron, but we will neglect that here.)

The volume of the nucleus (treating it as a sphere of radius R) is $V = \frac{4}{3}\pi R^3$.

Using $R = r_0 A^{1/3}$,

$$V = \frac{4}{3}\pi (r_0 A^{\frac{1}{3}})^3$$

$$= \frac{4}{3}\pi r_0^3 A$$

$$= A \times V_n$$

where V_n = the volume of one nucleon.

This gives the density as

$$\frac{A \times m_n}{A \times V_n} = \frac{m_n}{V_n}$$

This corresponds to the density of a single nucleon. The density of nuclear matter is therefore independent of which particular isotope we are considering. Nuclear matter has a density of approximately $2 \times 10^{17}\,\text{kg m}^{-3}$.

Essential Notes

This is an enormous density. If a table-tennis ball was made out of nuclear matter it would have a mass of about 8 million tonnes. If you squashed the Earth until it reached the density of nuclear matter, it would have a radius of only 200 m.

Example

Calculate the density of an alpha particle.

Answer

The alpha particle is a helium nucleus, so $A = 4$ and $Z = 2$. Its radius is $R = r_0 A^{1/3}$. $R = 1.2 \times 10^{-15} \times 4^{1/3} = 1.9 \times 10^{-15}$ m

The volume of the alpha particle is therefore $4/3\, \pi r^3 = 3 \times 10^{-44}\,\text{m}^{-3}$.

The mass of an alpha particle (2p and 2 neutrons) is 6.64×10^{-27} kg, so the density is mass / volume = $2.3 \times 10^{17}\,\text{kg m}^{-3}$.

3.8.1.6 Mass and energy

Mass difference and binding energy

The total mass of a nucleus is not simply the total mass of its constituents. When protons and neutrons are assembled into a nucleus, mass is lost. The mass of a nucleus of carbon-12 is 1.992×10^{-27} kg, but the combined mass of its 6 protons and 6 neutrons is 2.008×10^{-27} kg. The carbon nucleus weighs less than the sum of its parts. The missing mass is known as the **mass difference**.

If we wished to pull a nucleus apart, back into its separate nucleons, the missing mass would have to be replaced. In fact we would need to put in energy in order to pull the nucleus apart. This energy is known as the **binding energy**, and it is also the energy released when a nucleus is formed from its constituent nucleons. Einstein's theory of special relativity is used to connect the mass difference, Δm, to the binding energy, E:

$$E = \Delta m c^2$$

where c is the speed of light $= 2.998 \times 10^8\,\text{m s}^{-1}$.

Notes

When you use the equation $E = \Delta mc^2$, Δm must be in kg to give E in joules.

In the case of carbon-12 which has a mass difference of 1.6×10^{-29} kg, the binding energy is

$$E = 1.6 \times 10^{-29} \times 8.988 \times 10^{16} = 1.44 \times 10^{-12}\,\mathrm{J}$$

This is the amount of energy that would be needed to separate the 6 protons and 6 neutrons.

Large nuclei, like uranium-235, have a large mass difference and consequently have a large value of binding energy. However, this does not necessarily mean that they are stable. It is the **binding energy per nucleon** that tells us how much energy, on average, it will take to pull out each nucleon. The isotope with the highest value of binding energy per nucleon is iron-56, which is the most stable isotope.

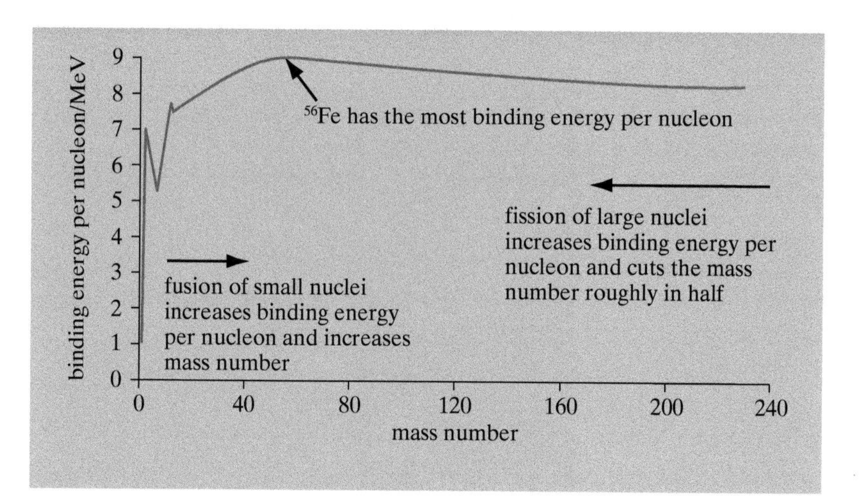

Fig 115 Binding energy per nucleon vs nucleon number

Essential Notes

Energy will be released by any nuclear reaction that results in products that have a higher binding energy per nucleon.

You can see from the graph in Fig 115 that very massive nuclei can increase their stability by splitting into two smaller isotopes. This is known as nuclear **fission** and certain isotopes of uranium and plutonium decay by spontaneously splitting into two smaller nuclei. A large amount of energy is released.

Energy is also released when two small nuclei merge to form a larger one. This is a nuclear **fusion**.

Energy from nuclear reactions

The energy released by nuclear reactions, such as alpha or beta decay, can also be explained by a change in mass. Radium-226 is an alpha particle emitter. It decays to radon, emitting an alpha particle with an energy of about 5 MeV:

$$^{226}_{88}\mathrm{Ra} \rightarrow ^{222}_{86}\mathrm{Rn} + \alpha$$

The mass of the radium nucleus is $3.753\,15 \times 10^{-25}$ kg

The mass of the radon nucleus is $3.686\,60 \times 10^{-25}$ kg

The mass of the alpha particle is 0.06646×10^{-25} kg

When we compare the original mass of the radium nucleus with the mass of the reaction products, we find that there is a mass difference of 9.00×10^{-30} kg. It is this mass which has been transferred to energy. Using the equation $E = \Delta mc^2$, we can calculate the energy released by the equation:

$$E = 9.00 \times 10^{-30} \times 8.988 \times 10^{16} = 8.1 \times 10^{-13}\ \text{J}$$

The usual SI units for mass and energy, kilograms and joules, are rather large for calculations on an atomic scale. We define smaller units, namely the **atomic mass unit, u,** and the **electron volt, eV.**

The atomic mass unit is defined as 1/12 of the mass of a carbon-12 atom. It is equal to 1.6605×10^{-27} kg. Using this unit a hydrogen atom has a mass of approximately 1 u. Accurate values are given in Table 6.

The electron volt is a more useful measure of energy on an atomic scale. 1 electron volt is equal to 1.602×10^{-19} J.

We can use Einstein's relation, $E = \Delta mc^2$, to calculate the energy equivalent of one atomic mass unit.

$$\begin{aligned} E = \Delta mc^2 &= 1.6605 \times 10^{-27} \times 8.988 \times 10^{16} \\ &= 1.492 \times 10^{-10}\ \text{J} \\ &= \frac{1.492 \times 10^{-10}}{1.602 \times 10^{-19}}\ \text{eV} \\ &= 9.315 \times 10^{8}\ \text{eV or } 931.5\ \text{MeV} \end{aligned}$$

This means that if 1 atomic mass unit of matter were entirely transferred to energy, it would release 931.5 MeV.

Essential Notes

Einstein's relation, $E = \Delta mc^2$, doesn't just apply to nuclear reactions. It links energy and mass changes for *all* physical systems. A moving object increases its mass as it accelerates. The relationship describes the energy needed to create particles in pair production, and the energy released by annihilation of matter and antimatter (see AS/A-Level Year 1, Section 3.2).

Essential Notes

You previously used electron volts to calculate energy changes in atoms in AS/A-Level Year 1, Section 3.2.

Notes

A quick way of finding the energy released in MeV is to multiply the mass difference in atomic mass units, u, by the factor 931.5.

Particle	Mass/kg	Mass/u	Energy equivalent/ MeV
Carbon-12 atom	1.99200×10^{-26}	12.0000	11175.6
Electron	9.11000×10^{-31}	5.48295×10^{-4}	0.51109
Proton	1.67208×10^{-27}	1.00728	938.080
Neutron	1.67438×10^{-27}	1.00867	939.374
Alpha particle	6.64250×10^{-27}	4.00151	3726.61

Table 7
Masses of some particles

Nuclear fission

Nuclear fission occurs in a few heavy nuclei, which decay by splitting into two roughly equal fragments (known as the fission products) and a few free neutrons and beta particles. Spontaneous fission is rare but fission can be induced in uranium-235 by allowing it to absorb another neutron. The resulting isotope, uranium-236, is very unstable. It decays by fission, often emitting two or three extra neutrons at high speed. The energy released in such a reaction can be calculated by finding the mass difference. The equation for a typical fission is

$$^{235}_{92}\text{U} + ^{1}_{0}\text{n} \rightarrow ^{134}_{54}\text{Xe} + ^{100}_{38}\text{Sr} + 2^{1}_{0}\text{n}$$

The total mass was originally

$$235.044\ \text{u}\ (^{235}_{92}\text{U}) + 1.009\ \text{u}\ (^{1}_{0}\text{n}) = 236.053\ \text{u}$$

After the fission the mass is

$$133.9054\,u\,(Xe) + 99.9354\,u\,(Sr) + 2 \times 1.009\,u\,(2n) = 235.859\,u$$

The mass difference is 0.194 u, giving an energy release of 180 MeV per fission. This is a huge energy output compared with even the most energetic chemical reactions.

Nuclear fusion

The energy that powers stars comes from nuclear fusion, the joining of two light nuclei to make a more massive one. Fusion reactions also release energy. For example, consider the fusion of two deuterium nuclei to make helium:

$$^{2}_{1}H + ^{2}_{1}H \rightarrow ^{3}_{2}He + ^{1}_{0}n$$

The mass of two deuterium nuclei is $2 \times 2.01410 = 4.02820$ u

The mass of helium-3 + a neutron is $3.01605 + 1.00867 = 4.02472$ u

This is a mass difference of 0.00348 u. We can convert this directly to energy in MeV:

$$\text{energy released} = 0.00348\text{ u} \times 931.5\text{ MeV} = 3.24\text{ MeV released}$$

The energy released from fusion reactions is enormous, typically around 10^{14} joules per tonne of fuel. This is even larger than the energy released from fission reactions, and yet there are no fusion power stations.

There are complex technical problems to be overcome before fusion can be used as a commercially viable energy source. The two nuclei have to be brought very close, within 10^{-15} m, before they will fuse. At this distance the electrostatic repulsion between two positively charged protons is huge. The protons have to be moving very quickly if they are to overcome this repulsion. This requires very high temperatures, of the order of 100 million degrees Celsius. It has proved difficult to contain and control material at such high temperatures. All the nuclear power stations built so far have been fission reactors.

Essential Notes

The fission products vary from fission to fission. A wide range of isotopes are produced. The energy released is typically 200 MeV per fission. Most of this is released instantly as kinetic energy of the fission products and neutrons, and energy of gamma rays. The remainder is released over a longer period of time as the unstable fission products decay, releasing beta-particles and neutrinos.

Essential Notes

Deuterium is an isotope of hydrogen.

Notes

Calculations of mass difference involve subtracting two numbers to find the small difference between them. It is essential to retain a large number of significant figures until the end of the calculation.

Essential Notes

At these very high temperatures all the electrons are stripped off their atoms. The material in an experimental fusion reactor is in the form of ionised gas, known as plasma.

3.8.1.7 Induced fission

Nuclear power stations produce about 15% of the electricity in the UK. This proportion is much higher in some other European countries. There are a number of different reactor designs but all the reactors use the energy released by nuclear fission, usually of uranium-235, to generate high-pressure steam. The steam drives turbines, which in turn drive electrical generators.

Natural uranium is a mixture of several different isotopes. Uranium-238 is by far the most common (99.28%), whilst uranium-235 is much less common (0.718%). Uranium-238 is only slightly fissile when exposed to very high-energy neutrons, whereas uranium-235 readily undergoes fission with low-energy (thermal) neutrons. So it is the relatively rare uranium-235 that is the useful fuel for nuclear reactors. Some reactors use fuel in which

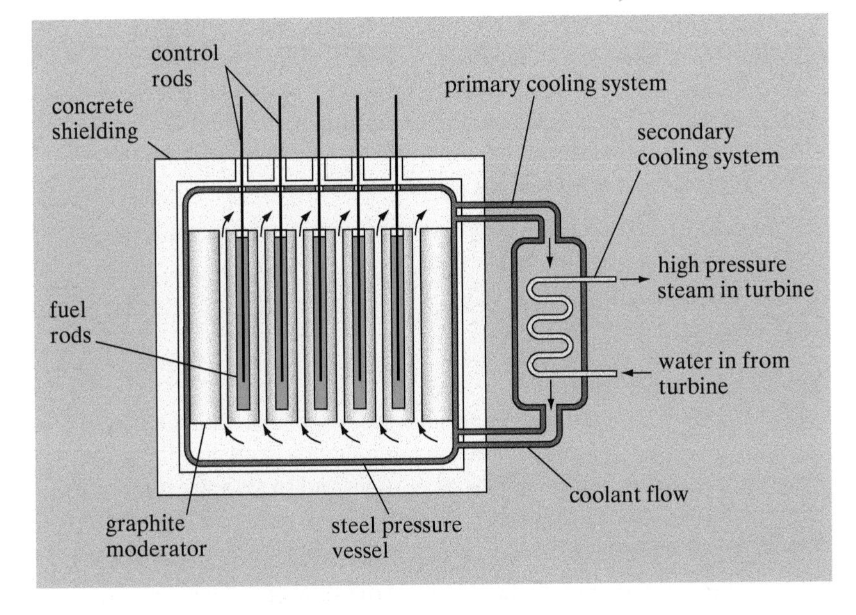

Fig 116
Gas-cooled nuclear reactor

the percentage of uranium-235 is increased. This is known as **enriched nuclear fuel**. Even so, the natural fission rate in uranium-235 is extremely slow. A practical nuclear reactor has to have a much greater power output. Bombarding the uranium-235 with neutrons can increase the fission rate. This is known as **induced fission**.

The chain reaction and critical mass

Each fission reaction releases extra neutrons. These can go on to induce further fission reactions. If all the neutrons that are released from each fission cause more fission reactions, then a chain reaction will ensue. A large amount of energy is then released in a very short time.

In practice not all the neutrons will cause further fission reactions. A neutron that is emitted from a fission reaction typically has an energy of 10 MeV, and is moving at a very high speed. A number of different things can happen to the neutron.

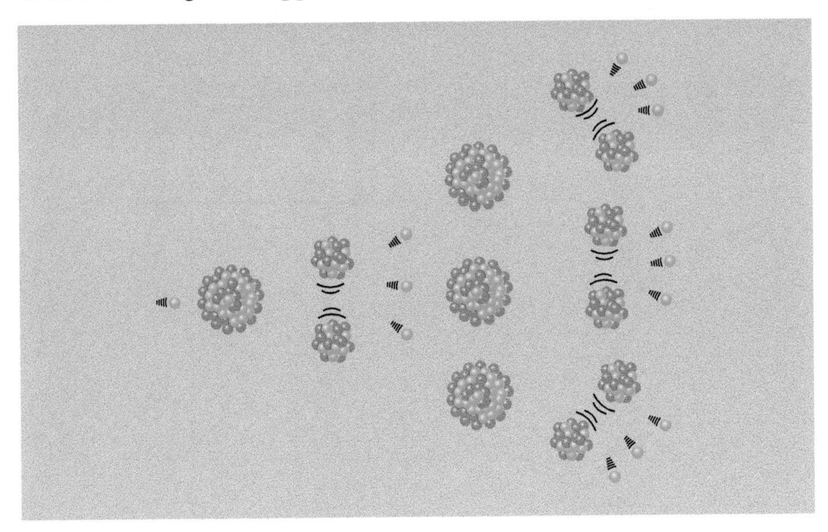

Fig 117
Chain reaction

One fission may release 3 neutrons. These can cause further fissions, releasing 9 neutrons, then 27 neutrons, and so on.

1. The neutron may leave the sample of uranium without causing any further reactions. In a small piece of uranium this is quite likely. In a larger piece of uranium, most of the neutrons will cause further fission. There is a **critical volume** for uranium below which a chain reaction cannot be sustained. For a spherical piece of uranium this corresponds to a **critical mass** of a few kilograms.
2. The neutron could be absorbed by uranium-238, or other nuclei, without causing any further fission.
3. The neutron could be absorbed by a nucleus of uranium-235 and cause another fission reaction.

Essential Notes

In an atomic bomb, two or more pieces of uranium (or plutonium), each less than the critical mass, are pushed together violently by a conventional chemical explosion. Once together their combined mass is greater than the critical mass, a rapid chain reaction takes place and there is an enormous explosion.

Moderation and control

The probability that uranium-235 will absorb a neutron depends on the neutron speed. There is a much greater chance of absorption if the neutron is travelling at low speeds. It is important to slow the neutrons down quickly, so that they can cause more fission reactions. A material known as a **moderator** is used to slow the neutrons down.

Essential Notes

Slow neutrons are often referred to as **thermal neutrons** because their energy is equal to the average kinetic energy of atoms of the medium. At 20 °C this is about 0.025 eV.

In a nuclear reactor the fuel is in the form of hundreds of narrow rods surrounded by the moderator. These fuel rods each contain less than the critical mass of fissionable uranium. Most of the emitted neutrons leave the fuel rod and collide with the atoms of the moderator, slowing down in the process, before reaching the next fuel rod at the right speed to cause further fission reactions. The material used for a moderator has to have a low mass number so that as much as possible of the neutron's kinetic energy is transferred at each collision. Suitable materials are water, since it contains low-mass hydrogen atoms, and graphite (a form of carbon).

Essential Notes

Imagine throwing a squash ball first at a football, then at a tennis ball. The football has a much larger mass and so will move off with a lower velocity. Since kinetic energy depends on velocity squared, the football will carry off less kinetic energy. The collision with the tennis ball will slow the squash ball down more. The most efficient transfer happens when both particles have the same mass.

If the reactor is to transfer energy at a steady rate then, on average, each fission reaction must lead to one more fission reaction. If there are too many fission reactions occurring, **control rods** are used to absorb neutrons. Control rods are made of a material like boron or cadmium,

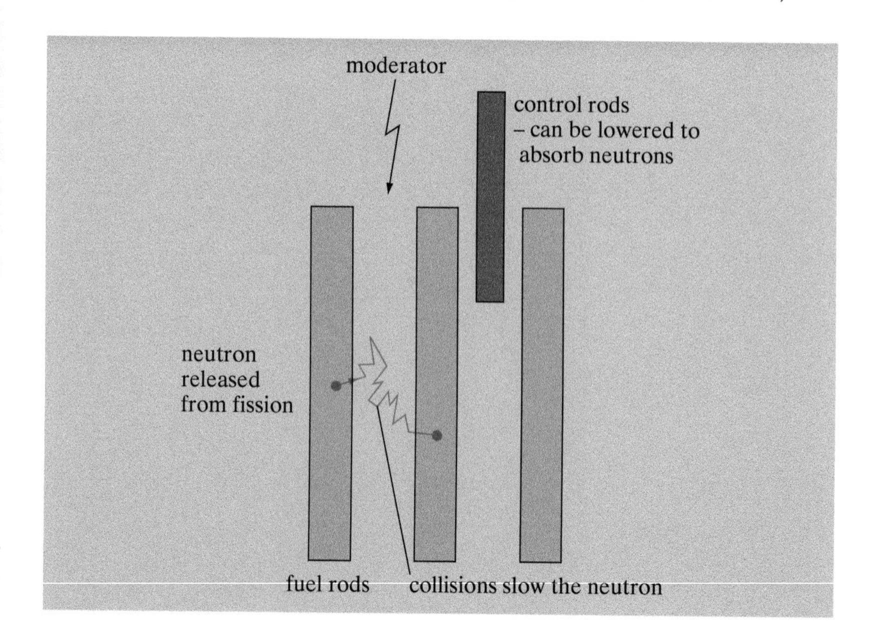

Fig 118
The roles of moderator and control rods in a nuclear reactor

which absorb neutrons well. These are lowered further into the reactor core to reduce the rate of fissioning, or raised to increase the rate.

Neutrons transfer energy to the moderator, which becomes very hot. This thermal energy is transferred to a **coolant**, often carbon dioxide or water, which in turn passes the energy on to a secondary coolant, water, that is turned to steam. The coolant has to be a fluid so that it can be pumped around the reactor core, and it should have a large specific heat capacity.

Essential Notes

Specific heat capacity is the energy needed to heat 1 kg of a substance by 1 °C (see page 32).

The neutrons released by the fission of uranium-235 are travelling at high speed, which reduces the probability that they will be absorbed by other uranium-235 nuclei and cause further fission. The moderator is there to slow the neutrons down to 'thermal' speeds, i.e. speeds typical of particles at that temperature. AGR (advanced gas cooled) reactors use graphite as a moderator. The mass of a graphite (carbon) atom is 12 times that of a neutron. Fig 119 shows a collision between the neutron and a carbon atom.

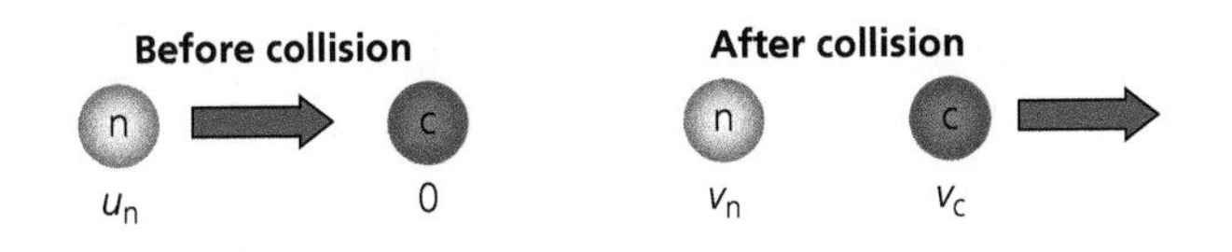

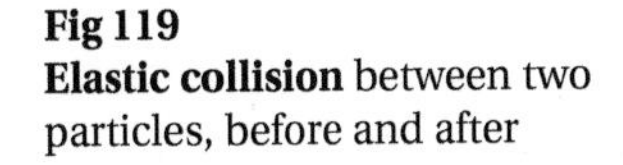
Fig 119
Elastic collision between two particles, before and after

Conservation of energy: $\frac{1}{2} m_n u_n^2 = \frac{1}{2} m_n v_n^2 + \frac{1}{2} m_c v_c^2$

Conservation of momentum: $m_n u_n = m_n v_n + m_c v_c$

These two equations can be combined to give the ratio of the neutron's kinetic energy loss to its original kinetic energy.

$KE_{\text{loss by neutron}} / KE_{\text{neutron}} = 4\ (m_c / m_n) / (1 + (m_c / m_n))^2$

When $m_c = 12\ m_n$, the ratio is $48 / 13^2 = 0.28$. So using graphite for a moderator means that at each collision the neutron passes on 0.28 of its kinetic energy to the moderator.

Example

(a) How many collisions must the neutron have before its kinetic energy has been reduced to 0.1% of its original value?

(b) Repeat the calculation in (a) supposing that the moderator was lithium rather than graphite.

(c) What would be the advantage in using lithium?

(d) Why isn't lithium used?

Answer

(a) After one collision the neutron has 0.72 of its original kinetic energy. After n collisions:

$(0.72)^n < 0.001$

$n \log (0.72) < \log (0.001)$

$n = -3 / -0.14$

$n = 21.4$. Therefore 22 collisions are needed.

(b) The neutron loses 0.75 of its original kinetic energy when it collides with the lithium nucleus:

$(0.25)^n < 0.001$

$n \log(0.25) < \log(0.001)$

$n = -3 / -0.60$

$n = 4.9$. Therefore five collisions are needed

(c) Fewer collisions are needed to bring the neutron down to thermal velocity, so the reactor could be much smaller.

(d) Lithium is extremely reactive, with water especially, not a sensible material to use in the core of a nuclear reactor. Lithium also has a low melting point, 180 °C.

3.8.1.8 Safety aspects

It is impossible for a nuclear explosion to occur in a power station. The fuel is not enriched enough and it would be impossible for a critical mass of fissionable uranium to come together. However, there is a danger from other explosions, which could lead to an escape of radioactive matter into the environment.

The worst nuclear accident occurred at Chernobyl in the Ukraine in 1986 when the reactor power got out of control. The coolant boiled and blew the huge concrete lid off the power station. To prevent accidents like this occurring, power stations have a number of safety features.

There is often a set of control rods held out of the reactor on large electromagnets. If the temperature gets too high, perhaps because of a power failure to the coolant fans, the control rods drop in and shut down the reactor. Power stations also have the capability to flood the reactor with nitrogen gas or, in the last resort, water. This would absorb any spare neutrons and stop all fission occurring.

The high neutron flux from the reactor, and the high levels of radiation from the fuel and the fission products, mean that substantial shielding is necessary. Thick steel shields and several metres of concrete surround the reactor.

A fuel rod of enriched uranium is slightly radioactive before it goes into the reactor, but it is much more dangerous after it has spent some time in the reactor. The reaction products left behind by the fission of uranium are neutron-rich, highly unstable and very radioactive. Products such as strontium-90 and caesium-137 are very dangerous to humans. The presence of highly radioactive fission products means that the fuel rod now has to be handled with remote apparatus. When the fuel rods are removed from the reactor they are dropped into a large pool of water where they are left to cool down. The high temperatures and the high levels of radioactivity are allowed to fall. Fuel rods are then transported for reprocessing where the unused uranium is recovered. The fission products are still very radioactive. These 'high-level' waste products are kept deep underground in geologically stable repositories.

Practical and mathematical skills

In both the AS and A-level papers at least 15% of marks will be allocated to the assessment of skills related to practical physics. A minimum of 40% of the marks will be allocated to assessing mathematical skills at level 2 and above. These practical and mathematical skills are likely to overlap to some extent, for example applying mathematical concepts to analysing given data and in plotting and interpretation of graphs.

The required practical activities assessed at AS are:

- Investigation into the variation of the frequency of stationary waves on a string with length, tension and mass per unit length of the string.
- Investigation of interference effects to include the Young's slit experiment and interference by a diffraction grating.
- Determination of *g* by a free-fall method.
- Determination of the Young modulus by a simple method.
- Determination of resistivity of a wire using a micrometer, ammeter and voltmeter.
- Investigation of the e.m.f. and internal resistance of electric cells and batteries by measuring the variation of the terminal p.d. of the cell with current in it.

The additional required practical activities assessed only at A-level are:

- Investigation into simple harmonic motion using a mass–spring system and a simple pendulum.
- Investigation of Boyle's (constant temperature) law and Charles's (constant pressure) law for a gas.
- Investigation of the charge and discharge of capacitors. Analysis techniques should include log–linear plotting leading to a determination of the time constant *RC*.
- Investigate how the force on a wire varies with flux density, current and length of wire using a top pan balance.
- Investigate, using a search coil and oscilloscope, the effect on magnetic flux linkage of varying the angle between a search coil and magnetic field direction.
- Investigation of the inverse-square law for gamma radiation.

Questions will assess the ability to understand in detail how to ensure that the use of instruments, equipment and techniques leads to results that are as accurate as possible. The list of apparatus and techniques is given in the specification.

Exam questions may require problem solving and application of scientific knowledge in practical contexts, including novel contexts.

Exam questions may also ask for critical comments on a given experimental method, conclusions from given observations or require the presentation of data in appropriate ways such as in tables or graphs. It will also be necessary to express numerical results to an appropriate precision with reference to uncertainties and errors, for example in thermometer readings.

The mathematical skills assessed are given in the specification.

Guidance on logarithms and logarithmic graphs

In your A-level examinations, you may be asked to construct or interpret graphs with a logarithmic scale. If you are studying A-level Mathematics you may have met logarithms already. This section tells you what you need to know for A-level Physics.

There are two main reasons for using logarithmic scales on graphs in physics. First, a graph with logarithmic scales enables you to show a much wider range of values, e.g. from 10 to 10 000, which would be impossible to show on a linear scale. Second, a logarithmic graph enables you to find an unknown power in an equation linking two variables, such as n in $y = x^n$.

The logarithm of a number to the base 10 is the power to which 10 must be raised to equal that number. Suppose that $x = 10^y$, we say that y is the logarithm of x to the base 10:

If $x = 10^y$ then $y = \log_{10}x$

For example, $100 = 10^2$ so the logarithm of 100 (to the base 10) is 2.

Suppose you had to plot a graph of the data shown in Table 1.

Table 1

Current/A	Power/W
1	1
10	100
100	10 000
1000	1 000 000
2000	4 000 000
3000	9 000 000

If you used an x-scale that allowed you to plot the first through to the last value on the x-axis (with 1 mm representing 1 amp), you would need rather large graph paper (3 m long!). It would need to be even larger to fit all the y values. Using logarithmic scales on both axes solves this problem.

Using your calculator to find the logarithms of the current and power values, you can add two new columns to the table – as shown in Table 2.

Table 2
Note the headings on the new columns – a logarithm is an index (power) and it has no units, even if the original quantity did have units. 'Log' is used to indicate 'logarithm to the base 10'.

Current/A	Power/W	log(current/A)	log(power/W)
1	1	0.00	0.00
10	100	1.00	2.00
100	10 000	2.00	4.00
1000	1 000 000	3.00	6.00
2000	4 000 000	3.30	6.60
3000	9 000 000	3.48	6.95

This data can now be plotted as a log–log graph.

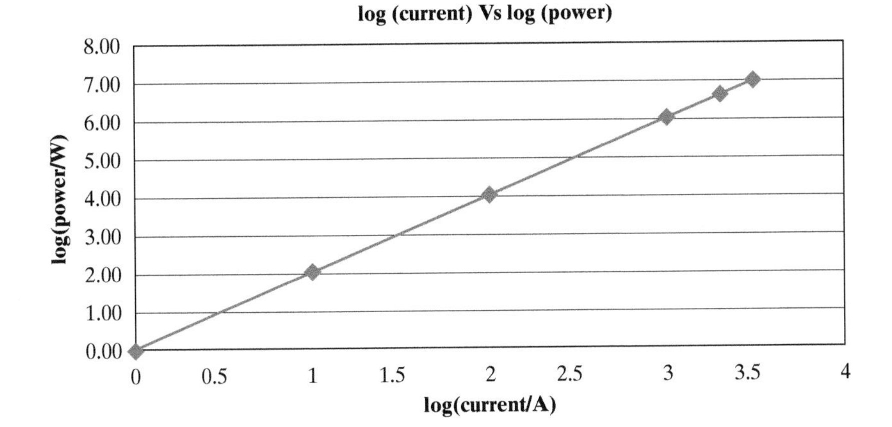

Logarithms transform multiplication to the process of addition:

$$\mathbf{\log_{10}(AB) = \log_{10}A + \log_{10}B}$$

This is because when we multiply two powers together we add the indices, e.g. $10^2 \times 10^3 = 10^5$.

Another feature of logarithms is that the process of raising a number to a certain power is transformed to multiplication:

$$\mathbf{\log_{10}(A^n) = n \times \log_{10}A}$$

This is because $(10^2)^3 = 10^2 \times 10^2 \times 10^2 = 10^6$, or $10^{2\times3}$.

These two properties of logarithms can be used to plot graphs to find unknown powers. For example, it is known that the force, F, between two bar magnets varies with the distance between them, x, according to the equation:

$$F = Ax^n$$

where A and n are unknown constants.

A graph of F against x would produce an exponential curve, so it is useful to use logarithms. Taking logs of both sides:

$$\log F = \log(Ax^n)$$

Using equation 1, $\log F = \log A + \log x^n$.

Using equation 2, $\log F = \log A + n \log x$.

Compare this to the equation of a straight line graph:

$$y = mx + c$$

If we plot $\log F$ on the y-axis and $\log x$ on the x-axis, the gradient (m) will give the power, n, and the intercept will be equal to $\log A$.

In Physics A-level studies you will often need to use logarithms to the base 10. However, it is sometimes useful to take logarithms to the base e, where e = 2.718. These are called natural logarithms. They are written in the form $\log_e x$, or $\ln x$ ('ln' means natural logarithm). The term $\ln x$ is the inverse function of e^x, so that $\ln e^x = x$.

Natural logarithms are used to plot exponential relationships, such as capacitor discharge and radioactive decay. For example, to find the time constant, CR, from capacitor discharge data giving the potential difference, V, at different times, t:

$$V = V_0 e^{\frac{-t}{CR}}$$

$$\frac{V}{V_0} = e^{\frac{-t}{CR}}$$

Taking natural logs of both sides gives:

$$\ln\frac{V}{V_0}=\frac{-t}{CR}$$

so that CR is given by:

$$CR=\frac{-t}{\ln\left(\frac{V}{V_0}\right)}$$

When you are dealing with exponential relationships, it is often useful to plot log–linear graphs, as shown in the example below.

Activity/Bq	Time/s
250	0
226	10
205	20
186	30
168	40
152	50
138	60
125	70

Example

Find the decay constant, and hence the half-life, of protactinium-234 from the data given. Assume that the data has already been corrected for background radiation.

The activity, A, at time t is given by $A = A_0e^{-\lambda t}$.

Taking logarithms to the base e:

$\ln A = \ln(A_0\, e^{-\lambda t})$

$\quad = \ln A_0 + \ln e^{-\lambda t}$

$\quad = \ln A_0 - \lambda t$

So plotting $\ln A$ on the y-axis against time t on the x-axis gives a straight line with a gradient of $-\lambda$.

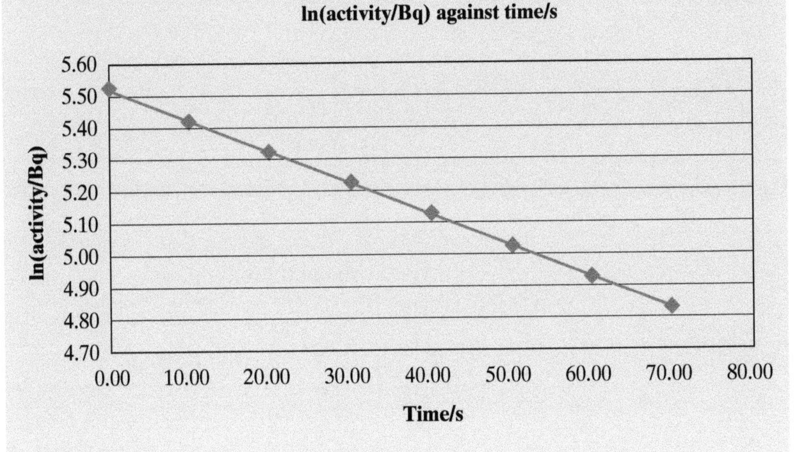

The gradient, $-\lambda$, is $-9.9 \times 10^{-3}\,s^{-1}$.

$$\text{Half-life} = \frac{\ln 2}{\lambda}$$

$$= \frac{0.693}{9.9\times10^{-3}\,s^{-1}}$$

$$= 70\ s$$

Data and formulae

FUNDAMENTAL CONSTANTS AND VALUES

Quantity	Symbol	Value	Units
speed of light in vacuo	c	3.00×10^{8}	$m\ s^{-1}$
permeability of free space	μ_0	$4\pi \times 10^{-7}$	$H\ m^{-1}$
permittivity of free space	ε_0	8.85×10^{-12}	$F\ m^{-1}$
magnitude of the charge of electron	e	1.60×10^{-19}	C
the Planck constant	h	6.63×10^{-34}	J s
gravitational constant	G	6.67×10^{-11}	$N\ m^2\ kg^{-2}$
the Avogadro constant	N_A	6.02×10^{23}	mol^{-1}
molar gas constant	R	8.31	$J\ K^{-1}\ mol^{-1}$
the Boltzmann constant	k	1.38×10^{-23}	$J\ K^{-1}$
the Stefan constant	σ	5.67×10^{-8}	$W\ m^{-2}\ K^{-4}$
the Wien constant	α	2.90×10^{-3}	m K
electron rest mass (equivalent to 5.5×10^{-4} u)	m_e	9.11×10^{-31}	kg
electron charge–mass ratio	e/m_e	1.76×10^{11}	$C\ kg^{-1}$
proton rest mass (equivalent to 1.00728 u)	m_p	1.67×10^{-27}	kg
proton charge–mass ratio	e/m_p	9.58×10^{7}	$C\ kg^{-1}$
neutron rest mass (equivalent to 1.00867 u)	m_n	1.67×10^{-27}	kg
gravitational field strength	g	9.81	$N\ kg^{-1}$
acceleration due to gravity	g	9.81	$m\ s^{-2}$
atomic mass unit (1 u is equivalent to 931.5 MeV)	u	1.661×10^{-27}	kg

ASTRONOMICAL DATA

Body	Mass/kg	Mean radius/m
Sun	1.99×10^{30}	6.96×10^{8}
Earth	5.97×10^{24}	6.37×10^{6}

GEOMETRICAL EQUATIONS

arc length $= r\theta$

circumference of circle $= 2\pi r$

area of circle $= \pi r^2$

curved surface area of cylinder $= 2\pi rh$

volume of cylinder $= \pi r^2 h$

area of sphere $= 4\pi r^2$

volume of sphere $= \frac{4}{3}\pi r^3$

ALGEBRAIC EQUATIONS

quadratic equation

$$x = \frac{\left(-b \pm \sqrt{\left(b^2 - 4ac\right)}\right)}{2a}$$

MECHANICS

force $F = \frac{\Delta(mv)}{\Delta t}$

impulse $F\Delta t = \Delta(mv)$

CIRCULAR MOTION

angular velocity $\omega = \frac{v}{r}$

$\omega = 2\pi f$

centripetal acceleration $a = \frac{v^2}{r} = \omega^2 r$

centripetal force $F = \frac{mv^2}{r} = m\omega^2 r$

decay of charge $Q = Q_0(1 - e^{-t/RC})$

SIMPLE HARMONIC MOTION

acceleration $a = -(2\pi f)^2 x$

displacement $x = A\cos(\omega t)$

speed $v = \pm\omega\sqrt{A^2 - x^2}$

maximum speed $v_{max} = \omega A$

maximum acceleration $a_{max} = (\omega)^2 A$

for a mass-spring system $T = 2\pi\sqrt{\frac{m}{k}}$

for a simple pendulum $T = 2\pi\sqrt{\frac{l}{g}}$

GRAVITATIONAL FIELDS

force between two masses $F = -\frac{G m_1 m_2}{r^2}$

gravitational field strength $g = \frac{F}{m}$

magnitude of gravitational field strength in radial field $g = \frac{GM}{r^2}$

gravitational potential $\Delta W = m\Delta V$

$V = -\frac{GM}{r}$

$g = -\frac{\Delta V}{\Delta r}$

ELECTRIC FIELDS AND CAPACITORS

force between two point charges $F = \frac{1}{4\pi\varepsilon_0}\frac{Q_1 Q_2}{r^2}$

force on a charge $F = EQ$

field strength for a uniform field $E = \frac{V}{d}$

field strength for a radial field $E = \frac{Q}{4\pi\varepsilon_0 r^2}$

work done $\Delta W = Q\Delta V$

electric potential $V = \frac{1}{4\pi\varepsilon_0}\frac{Q}{r}$

capacitance $C = \frac{Q}{V}$

$C = \frac{A\varepsilon_0\varepsilon_r}{d}$

decay of charge $Q = Q_0 e^{-t/RC}$

time constant RC

capacitor energy stored $E = \frac{1}{2}QV = \frac{1}{2}CV^2 = \frac{1}{2}\frac{Q^2}{C}$

MAGNETIC FIELDS

force on a current $F = BIl$

force on a moving charge $F = BQv$

magnetic flux $\Phi = BA$

magnetic flux linkage $\Phi = BAN$

induced emf $\varepsilon = N\frac{\Delta\Phi}{\Delta t}$

emf induced in a rotating coil $N\Phi = BAN\cos\theta$

$\varepsilon = BAN\omega\sin\omega t$

transformer equations $\frac{N_s}{N_p} = \frac{V_s}{V_p}$

efficiency $= \frac{I_s V_s}{I_p V_p}$

NUCLEAR PHYSICS

the inverse square law for γ radiation $I = \frac{k}{x^2}$

radioactive decay $\frac{\Delta N}{\Delta t} = -\lambda N$

$N = N_0 e^{-\lambda t}$

activity $A = \lambda N$

half life $T_{½} = \frac{\ln 2}{\lambda}$

nuclear radius $R = R_0 A^{1/3}$

energy–mass equation $E = mc^2$

THERMAL PHYSICS

energy to change temperature $Q = mc\,\Delta T$

energy to change state $Q = ml$

gas law $pV = nRT$

$pV = NkT$

kinetic theory model $pV = \frac{1}{3}Nm\left(c_{rms}\right)^2$

kinetic energy of gas molecule $\frac{1}{2}m\left(c_{rms}\right)^2 = \frac{3}{2}kT = \frac{3RT}{2N_A}$

Practice exam-style questions

Section A: Multiple choice

For each question there are four responses. Select the response which you think is the most appropriate answer to a question. One mark for each correct answer.

1 Which one of these statements about a pendulum oscillating with simple harmonic motion is false?

A The frequency depends on the length of the pendulum.

B The maximum speed occurs at the same time as the minimum acceleration.

C The velocity is always in the opposite direction to the displacement.

D The acceleration is always in the opposite direction to the displacement.

2 A satellite is in a stable circular orbit around the Earth. Which of these statements is true?

A There is no resultant force on the satellite.

B The satellite is accelerating at a constant rate.

C The weight of the satellite is balanced by the centrifugal force.

D The satellite is weightless.

3 Planet Y has a mass, M, and a radius, R. The gravitational field strength at the surface of planet Y is g. Planet Z has the same density as planet Y but a radius of $2R$. The gravitational field strength at the surface of planet Z would be:

A $2g$

B g

C $4g$

D $8g$

4 This question is about the motion of charged particles in electric and magnetic fields. Which of the following alternatives is true (assume the particle is initially at rest)?

	Particle	Field	Particle trajectory
A	Electron	Uniform electric field from left to right	Accelerates from left to right
B	Electron	Uniform magnetic field from left to right	Accelerates from left to right
C	Positron	Uniform electric field from left to right	Accelerates from left to right
D	Positron	Uniform magnetic field from left to right	Accelerates from left to right

5 A 10 mF capacitor is charged to a potential difference of 10 V. It is then discharged through a 10 kΩ resistor for 1 minute. The average current during that time will be:

A 0.045 A

B 0.92 mA

C 0.055 mA

D 0.75 mA

6 A radioisotope is to be used for a medical diagnosis of kidney disease. Small amounts of the radioisotope will be added to a drink and the radiation emanating from the kidneys will be monitored. Which of these isotopes would be most suitable?

	Radiation emitted	Decay constant/s^{-1}
A	Beta	2.4×10^{-5}
B	Beta	2.4×10^{-3}
C	Gamma	2.4×10^{-5}
D	Gamma	2.4×10^{-7}

7 Which of the following set of material properties would be appropriate for a material to be used as a moderator in a nuclear power station?

A	Readily absorbs neutrons	High melting point	High atomic mass number
B	Does not readily absorb neutrons	High melting point	Low atomic mass number
C	Readily absorbs neutrons	Low melting point	Low atomic mass number
D	Does not readily absorb neutrons	High specific heat capacity	High atomic mass number

8 Estimate how many molecules of helium there are in a party balloon. (You could use the ideal gas equation, $pV = nRT$ to help.)

A 2.5×10^{22}

B 1.7×10^{20}

C 2.5×10^{18}

D 1.7×10^{28}

9 A coil of area A is inclined so that the normal to plane of the coil is at an angle α to a magnetic field of flux density B. If the coil has N turns, the flux linkage through the coil is given by

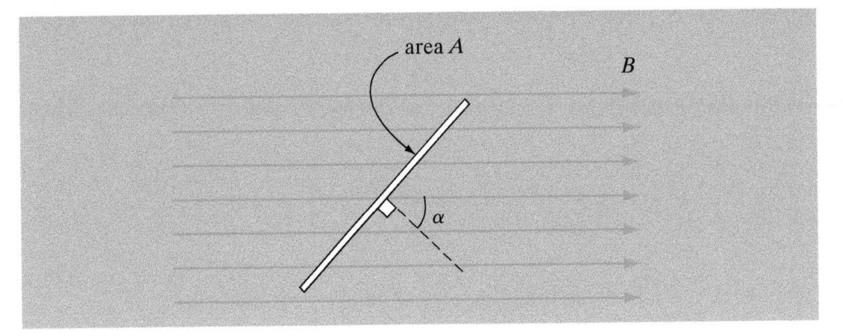

A $BA \cos \alpha$

B $BAN \sin \alpha$

C $BAN \cos \alpha$

D BAN

10 A step-down transformer is used to provide a 12 V supply when connected to the mains supply (230 V) in a laboratory. The transformer is 80% efficient and the current through the primary coil is 0.5 A. Which row in the table is correct?

	Current through the secondary /A	**Turns on primary**	**Turns on secondary**
A	3.8	26	500
B	0.2	26	500
C	3.8	500	26
D	7.7	500	26

Section B: Structured response

1 A cathode ray tube uses a high voltage anode to accelerate a beam of electrons across a vacuum tube. The potential difference between the cathode and the anode is 2000 V.

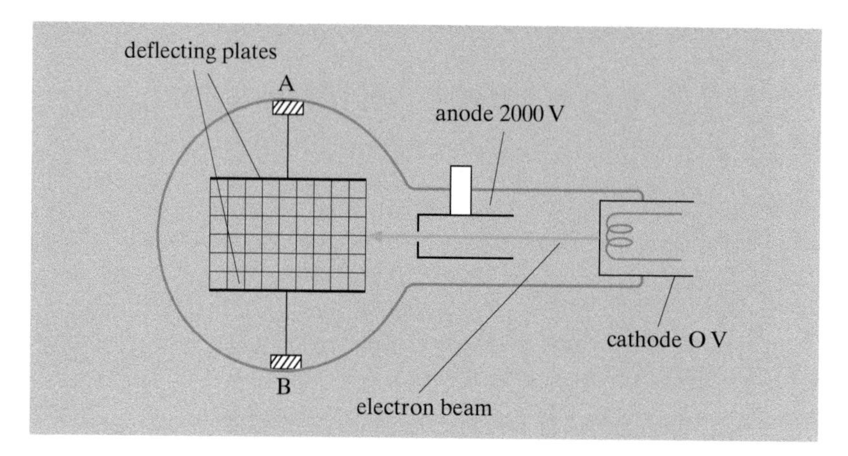

(a) Calculate the energy gained by an electron as it accelerates from the cathode to the anode. Give your answer in joules.

____

____ 2 marks

(b) Show that the velocity of an electron as it reaches the anode is 2.7×10^7 m s^{-1}.

____

____ 2 marks

(c) The deflecting plates are 5 cm apart. A potential of +3 kV is applied to the top plate. The bottom plate is connected to 0 V.

(i) Describe the path of the electron as it travels between the plates.

____

____ 2 marks

(ii) Calculate the electric field strength between the deflection plates.

2 marks

(iii) Calculate the acceleration of the electron as it moves between the deflection plates.

2 marks

Total Marks: 10

2 A child on a playground swing is pulled back and released, so that she oscillates with a natural frequency of 0.4 Hz.

(a) Explain what is meant by the natural frequency of a system.

2 marks

(b) If the child and swing behave like a simple pendulum, calculate the length of the swing.

2 marks

(c) The child is pulled back so that the initial displacement is 0.5 m, and then released. Calculate the displacement after 1 second has elapsed. (You may neglect the effect of damping.)

2 marks

(d) The oscillations are damped.

(i) Explain what is meant by 'damped' oscillations.

2 marks

(ii) Sketch a graph showing three full oscillations.

4 marks

Total Marks: 12

3 A bar magnet is dropped through a coil of wire that is connected to an oscilloscope. The output of the oscilloscope is shown below.

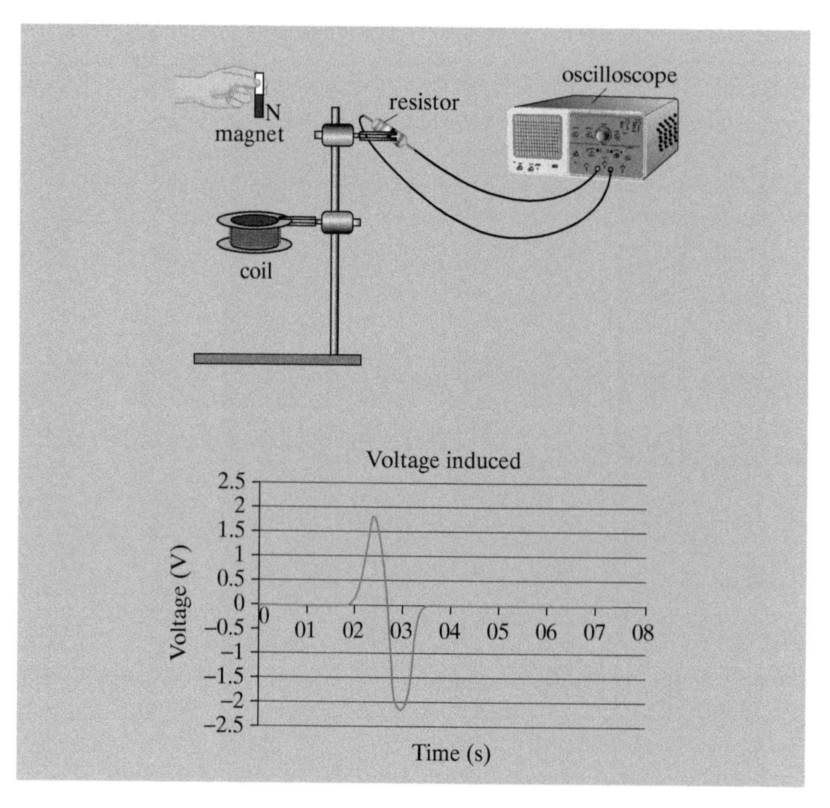

(a) Explain why a voltage is induced in the coil.

____

____

____ 2 marks

(b) State and explain what would happen to the oscilloscope display if the magnet was dropped from a greater height.

____

____

____ 2 marks

(c) How would the trace change if the magnet was turned, so that the south pole entered the coil first?

____

____ 1 mark

Total Marks: 5

4

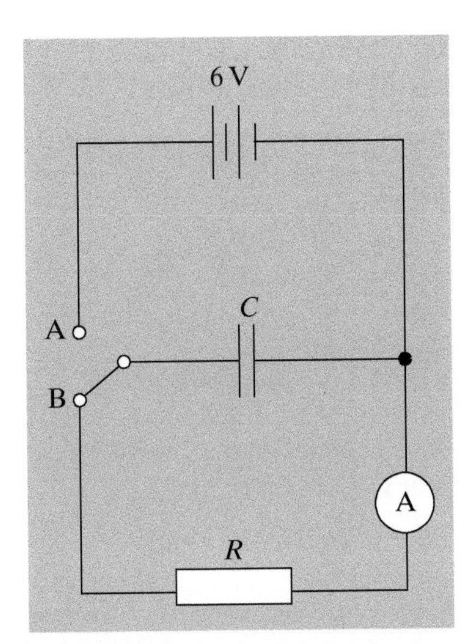

A capacitor is initially connected to the 6 V battery by moving the switch to A. At time $t = 0$, the switch is moved to B and the capacitor is discharged through the resistor. The capacitor has a capacitance of 100 mF, and the resistor has a resistance of 1 kΩ.

(a) Explain what is meant by the 'time constant' of this circuit.

____

____ 1 mark

(b) Calculate the time constant for this circuit.

____

____ 2 marks

(c) What will the initial current reading on the ammeter be when the switch is moved from A to B?

____

____ 1 mark

(d) How long will it take for the ammeter reading to drop to 1 mA?

____

____

____ 3 marks

Total Marks: 7

5 Cassini-Huygens is an unmanned spacecraft which is studying Saturn and its moons. It was launched from Earth in 1997 and reached the Saturn moon, Titan, in 2005.

The spacecraft is powered by a radioactive thermal generator which uses heat generated by radioactive decay to generate electricity. The radioisotope plutonium-238 is used.

The table shows the properties of three isotopes of plutonium.

	Pu-238	Pu-239	Pu-241
Half-life (in years)	87.74	24 110	14.4
Specific activity (GBq/gram)	640	0.063	104
Principal decay mode	alpha	alpha	beta
Decay energy (MeV)	5.593	5.244	0.021
Radiological hazards	alpha, weak gamma	alpha, weak gamma	beta, weak gamma

(a) What is meant by the **activity** of a radioactive source?

1 mark

(b) Show that a power of 573 W is generated by the alpha emissions of 1 kg of plutonium-238.

(Use data from the table above.)

3 marks

(c) The electricity generation process is 6.7% efficent. What mass of plutonium-238 is needed to generate 600 W of electrical power?

3 marks

(d) The Cassini mission was planned to last for 11 years. If the spacecraft needed 600 W at the end of the mission, what mass of plutonium-238 was needed at the beginning of the mission?

3 marks

(e) Explain why plutonium-238 was preferred to the other isotopes shown in the table above.

2 marks

(f) There were a number of protests against the Cassini mission, principally due to the use of radioactive power supplies. Explain why the protestors may have been concerned.

2 marks

(g) Suggest why a radioactive power source was preferred to using solar panels to power the spacecraft.

2 marks

Total Marks: 16

6 In October 2009, a large helium balloon belonging to amateur scientist, Richard Heene, floated off from his garden in Colorado, USA. At the same time, Heene's 6-year-old son went missing. Panic ensued until the boy was found safe and sound a few hours later.

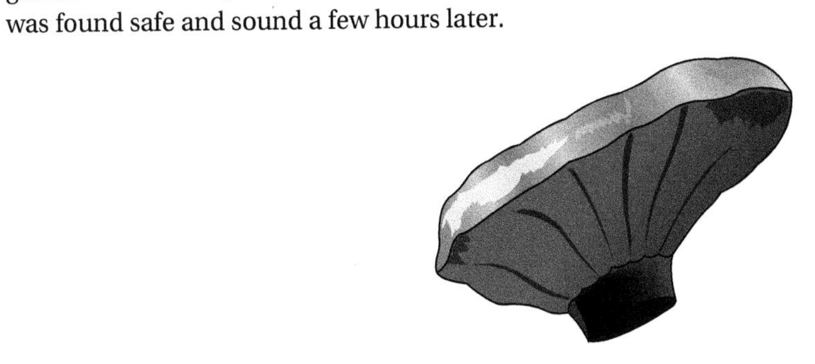

(a) The volume of the balloon was 28.3 m^3. If the temperature was 20 °C and atmospheric pressure was 1.01×10^5 Pa, calculate the mass of helium in the balloon. (The molar mass of helium is 4.0×10^{-3} kg.)

3 marks

(b) Suppose the balloon rose to a height of 5 km, where the air pressure was 0.5×10^5 Pa and the temperature was −40 °C. Calculate the new volume of the balloon. State any assumptions that you have made.

3 marks

(c) Any helium balloon has a maximum height that it can reach. Suggest why there is a maximum height.

1 mark

Total Marks: 7

7 (a) Explain what is meant by the **mass difference** and the **binding energy** of a nucleus.

(i) Mass difference

(ii) Binding energy

2 marks

(b) The graph below shows how the binding energy per nucleon varies with nucleon number.

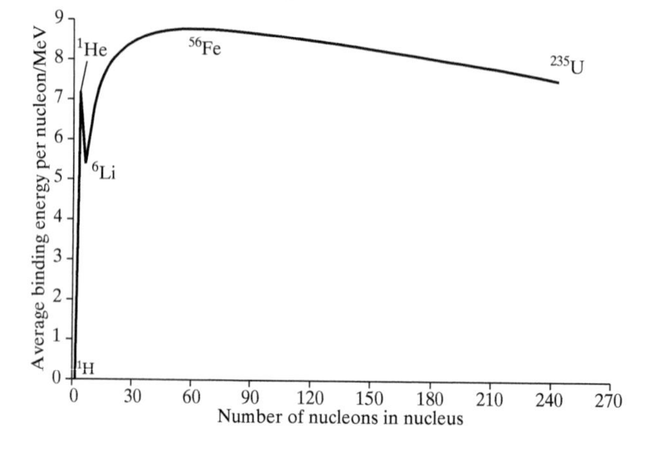

(i) The mass of helium-4 is 4.00151 u. Show that the binding energy per nucleon for $^{4}_{2}He$ plotted on the graph, 7.1 MeV, is correct. (You will need to use other values from the data sheet.)

3 marks

(ii) Use the graph to explain why energy is released by the nuclear fission of a uranium-235 nucleus.

2 marks

(c) In a nuclear reactor, uranium-235 absorbs a neutron to become uranium-236, which decays by fission:

$$^{235}_{92}U + ^{1}_{0}n \rightarrow ^{236}_{92}U \rightarrow ^{94}_{40}Zr + ^{140}_{56}Ba + 2^{1}_{0}n \text{ (+ beta particles)}$$

Calculate the energy released by this fission reaction. Give your answer in joules.

Mass of U-235 = 235.044 u

Mass of Zr-94 = 93.906 u

Mass of Ba-56 = 139.91 u

(See the data sheet for other values needed.)

3 marks

Total Marks: 10

8 During 1 hour of exercise in a gym, a person may lose 500 g of water through sweating. Sweating is a vital part of the body's mechanism for maintain a steady temperature. It is important to replace the lost water by drinking water.

The latent heat of vaporisation of sweat is 2.4×10^{6} J kg^{-1}.

(a) Explain what is meant by the **latent heat of vaporisation** of a liquid.

2 marks

(b) Explain how sweating helps to keep a person cool.

2 marks

(c) Calculate the average rate of heat loss through sweating for an hour's exercise in the gym.

3 marks

Total Marks: 7

9 An experiment involving a radioactive isotope is being carried out in a physics laboratory. The isotope is a solid metal of activity 1 MBq and half-life of 8 days. The isotope is thought to emit alpha and gamma radiation, but there may be some beta radiation from daughter nuclei.

Imagine that you are responsible for health and safety in the laboratory.

(a) Write brief working guidelines to ensure the health and safety of those working with the experiment. The guidelines should recommend safe working practices, and explain the reasoning behind each recommendation. Write your guidelines in clear, concise English. 6 marks

(b) You decide to measure how the radiation count rate varies with distance, d, from the radioisotope. The count rate, I, should follow the general formula:

$$I = I_0 d^n$$

Explain how you can find the value of n by plotting a log–log graph. 2 marks

(c) Use the data given here to plot a graph of log (count rate) versus log (distance). 4 marks

Count rate per second	Distance/cm
516	2
237	4
45	6
27	8
22	10
15	15
12	20

Background cps
10
9
9
5
8
8
9

(d) Explain whether your graph supports the suggestion that the isotope emits gamma and alpha radiation. 2 marks

10 A manometer is used to measure pressure differences between two gases. It consists of a U-shaped tube, partly filled with liquid.

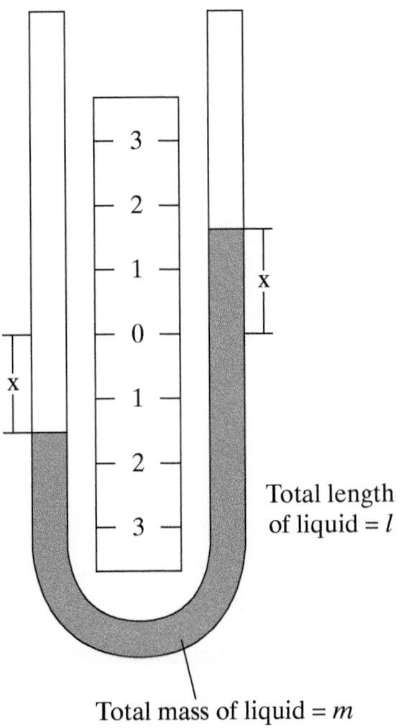

When the pressure of gas on one side is briefly increased, the water is forced to oscillate from one side of the U-tube to the other.

(a) When the liquid is at the position shown in the diagram, how much extra mass of liquid is in the right-hand side of the U-tube, compared to the left-hand side? 1 mark

____

(b) Look at this expression for the restoring force on the liquid:

$$F = (2x\ /\ l)\ Mg$$

and explain why this shows that the motion is simple harmonic. 2 marks

____

____

____

(c) Write down an expression for the acceleration of the water in the tube and compare it to the standard SHM equation for acceleration. Use this comparison to find the frequency, f, of the oscillations. 4 marks

____

____

____

(d) Describe how you would carry out an experiment to check the relationship, $f = \pi \sqrt{(2l / g)}$.

You should include any steps that you could take to improve accuracy and precision.

Explain what graph you would plot and how you could use it to find a value for g. 4 marks

Answers

Section A

1 mark for each correct answer.

Question	Answer	Comment
1	C	
2	B	
3	A	
4	C	For circular motion, the resultant force is towards the centre and at right angles to the velocity, so no work is done.
5	D	The initial charge on the capacitor = $10 \times 10^{-3} \times 10 = 0.1$ C. After 60 seconds of discharge $Q = 0.1\ e^{-60/(10\,000 \times 0.01)} = 0.055$ C Therefore the charge that has flowed = 0.1 – 0.055 = 0.045 C. This happened in 60 seconds so the average current is 0.045 / 60 = 0.75 mA.
6	C	Beta radiation would not penetrate the body sufficiently to be detected. C has a half-life of 8 hours, any longer than this would mean that the patient was radioactive for too long. The half-life has to be long enough to enable the isotope to be transported to the hospital but short enough to prevent the patient from irradiating their friends and family for days.
7	B	A moderator needs to slow neutrons (so needs an atomic mass number similar to the mass of a neutron, i.e., as low as possible). It must not absorb neutrons as it would stop the reaction. It must have a high melting point as the core of the reactor gets very hot.
8	A	$pV = nRT$ Volume is about 1 litre = $0.001\ m^3$ $P \approx 1 \times 10^5$ Pa $T = 290$ K $R = 8.31\ J\ mol^{-1}\ K^{-1}$. So $n = 0.04$ mol = 2.5×10^{22}
9	C	The flux linkage is AN times the component of B perpendicular to A.
10	D	Step-down the voltage by a factor 12/230 = 1/19, so current goes up by 19 times to 9.5 A, but transformer only 80% efficient so $I = 7.6$ A. Need a ratio of 19 for turns, so 500 primary and 26 secondary is correct.

Section B

Question	Answer		Marks
1 (a)(i)	Energy = $eV = 1.6 \times 10^{-19} \times 2000$ $= 3.2 \times 10^{-16}$ J	(1) (1)	2
1 (b)	Energy = $\frac{1}{2}mv^2$; so $v = \sqrt{(2E/m)}$ $= 2.65 \times 10^7\ m\ s^{-1}$ (In 'show that' questions, always give the next significant figure to prove you actually did the calculation.)	(1) (1)	2

Question	Answer		Marks
1 (c)(i)	Curving upwards in a parabolic (not circular) path.	(1) (1)	2
1 (c)(ii)	Field strength is 3000/0.05 = 60 kV m^{-1} or kN C^{-1} (1 mark for correct unit.)	(2)	2
1 (c)(iii)	Acceleration = $F/m = eE/m = 1.6 \times 10^{-19} \times 60\,000/9.11 \times 10^{-31}$ $= 1.1 \times 10^{16}$ m s^{-2}	(1) (1)	2
			Total 10
2 (a)	The number of free/unforced oscillations per second. Free oscillations occur when an object is disturbed from equilibrium and released.	(1) (1)	2
2 (b)	$T = 2\pi\,\sqrt{(l/g)}$; so $l = T^2g/4\pi^2$ = 1.6 m (T = 1/0.4 = 2.5 s)	(1) (1)	2
2 (c)	$x = A \cos \omega t = 0.5 \cos (2\pi ft) = 0.5 \cos\ (2\pi \times 0.4 \times 1)$ $= -0.50$ m (Must have −ve sign for second mark.)	(1) (1)	2
2 (d)(i)	Damped oscillations occur when energy is transferred from the oscillating system/work is done against resistive forces. The amplitude of the oscillations decreases with time.	(1) (1)	2
2 (d)(ii)	x and y axes labelled and suitable scales. Period = 0.4 s for all 3 cycles. Initial amplitude = 0.5 m, subsequent amplitude gets smaller.	(1) (1) (1) (1)	4
			Total 12
3 (a)	As the magnet drops through the coil, the flux linkage changes. Faraday's law states that e.m.f. is equal to the rate of change of flux linked.	(1) (1)	2
3 (b)	Induced voltage would be greater/trace has greater amplitude. Magnet moving quicker, so rate of change of flux is greater/less time to pass through coil.	(1) (1)	2
3 (c)	The trace would be inverted/go negative first.	(1)	1
			Total 5
4 (a)	The time constant is the time taken for the voltage/current/charge to drop to 1/e of its initial value.	(1)	1
4 (b)	$CR = 100 \times 10^{-3} \times 1 \times 10^3$ = 100 s	(1) (1)	2
4 (c)	$I = V/R$ = 6/1000 = 0.006 A or 6 mA	(1)	1
4 (d)	$I = I_0 \exp(-t/CR)$; so $I/I_0 = 1/6$ and $I_0/I = 6$ $t = CR \ln 6$ = 180 s	(1) (1) (1)	3
			Total 7

Question	Answer		Marks
5 (a)	Activity is the number of emissions per second.	(1)	1
5 (b)	Activity of 1 kg is 640 GBq × 1000 = 640 × 10^{12} Bq. Each alpha particle is emitted with energy 5.593 MeV = 5.593 × 10^6 × 1.6 × 10^{-19} J = 8.95 × 10^{-13} J So power (= energy/sec) = 8.95 × 10^{-13} J × 640 × 10^{12} s^{-1} = 573 W.	(1) (1) (1)	3
5 (c)	1 kg generates 573 W, but electricity generation only 6.7% efficient so electrical power = 0.067 × 573 = 38.4 W. So 1 W of electrical power needs 1000/38.4 = 26 g plutonium. And 600 W needs 15.6 kg.	(1) (1) (1)	3
5 (d)	Since $N = N_0 e^{-\lambda t}$ and mass $m \propto N$, $m = m_0 e^{-\lambda t}$ $\lambda = \ln 2/T_{1/2} = 0.693/87.74 = 7.9 \times 10^{-3}$ years^{-1} $\lambda t = 7.9 \times 10^{-3}$ years^{-1} × 11 years = 0.087 $m_0 = m e^{\lambda t} = 15.6 e^{0.087} = 17$ kg	 (1) (1) (1)	3
5 (e)	Longer half-life needed for long mission into space. Alpha emitter so easier to shield. Alpha emitter so all energy transferred close to power source. High power density/lots of energy. (Any 2 points)	(2)	2
5 (f)	Danger of radioactive/plutonium contamination if rocket blew up/crashed. Radioactive materials can cause increased cancer risk/ plutonium is toxic.	(1) (1)	2
5 (g)	Sun's energy is too dilute/inverse-square law for Sun's radiation means that solar radiation not intense enough at the distance of Saturn Solar panels would have to be very large/too heavy for lift-off.	(1) (1)	2
			Total 16
6 (a)	$pV = nRT$, so $n = PV/RT$ = (1.01 × 10^5 × 28.3)/(8.31 × 293) = 1170 moles Since 1 mole of helium has a mass of 4 × 10^{-3} kg, the balloon contains 1174 × 4 × 10^{-3} = 4.7 kg of helium.	(1) (1) (1)	3
6 (b)	pV/T is constant, so $V_2 = p_1V_1T_2/T_1p_2$ = (1.01 × 10^5 × 28.3 × 233)/(293 × 0.5 × 10^5) = 46 m^3 This assumes that the gas behaves as an ideal gas/ no helium escapes.	(1) (1) (1)	3
6 (c)	Pressure of atmosphere gets less with height, so helium balloon keeps expanding until it bursts.	(1)	1
			Total 7

Question	Answer		Marks
7 (a)(i)	Mass difference is the difference between the mass of a nucleus and the total mass of its constituent nucleons (protons and neutrons).	(1)	1
7 (a)(ii)	Binding energy is the energy required to split a nucleus into its constituent nucleons, or the energy released when the nucleus is formed from its constituents.	(1)	1
7 (b)(i)	Mass difference = $2m_p + 2m_n - 4.00151$ $= (2 \times 1.00728) + (2 \times 1.00867) - 4.00151 = 0.030$ u Binding energy = 0.030×931.3 MeV = 28.3 MeV Binding energy per nucleon = 28.3/4 = 7.08 MeV/nucleon	(1) (1) (1)	3
7 (b)(ii)	U-235 has binding energy per nucleon of about 7.5 MeV/nucleon. Fission produces two smaller nuclei; these have a larger binding energy per nucleo (about 8.5 MeV/nucleon). This means there is an overall mass loss in the fission, and so energy is released (as kinetic energy of the fission products).	(1) (1)	2
7 (c)	Original mass = 235.044 u + 1.009 u = 236.053 u Final mass = 93.906 u + 139.91 u + (2×1.009 u) = 235.836 u Mass difference = 0.217 u, giving an energy of $0.217 \times 931.3 = 202$ MeV $= 202 \times 10^6 \times 1.6 \times 10^{-19} = 3.23 \times 10^{-11}$ J	(1) (1) (1)	3
			Total 10
8 (a)	The energy required to vaporise 1 kg of liquid with no increase in temperature	(1) (1)	2
8 (b)	Water in sweat needs energy to evaporate/latent heat of vaporisation. This energy comes from the body's heat, causing it to cool down.	(1) (1)	2
8 (c)	Heat loss in 1 hour $Q = ml$ $= 0.5 \times 2.4 \times 10^6 = 1.2 \times 10^6$ J Average rate of heat loss = 1.2×10^6 J/3600 s = 330 W.	(1) (1) (1)	3
			Total 7
9 (a)	• Keep your distance from the radioisotope. Alpha and beta radiation have limited range in the air, and gamma radiation diminishes according to the inverse square law, so keeping your distance reduces your exposure. • Work as quickly as possible. The longer you are exposed, the greater the dose and the greater the risk. • Make sure that the isotope is appropriately stored when not in use. A lead-lined box and a locked storage facility should be used. • Do not handle source. Use gloves to avoid contamination. Use long-handled tongs when necessary. • Use of personal radiation dose monitoring is important when the activity of source is high or exposure times are long. This could be a film badge or a TLD dosimeter.		6

9 (b)	Taking logarithms of both sides: $\log I = \log I_0 + n \log d$ Compare this with: $y = mx + c$	2
9 (c)	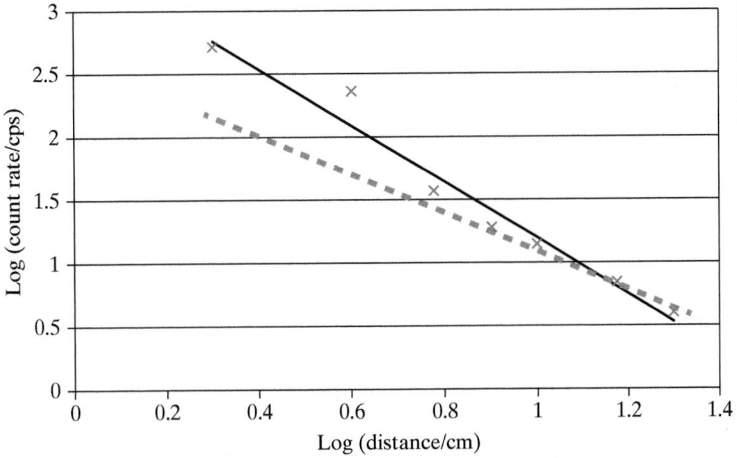	4
9 (d)	The best-fit line gives $n = -2.2$, but if we ignore the first two readings (where the alpha particles are being detected), much closer to $n = -2$ which we would expect (inverse square law). So it does suggest that alpha and gamma are being emitted.	2
		Total 14
10 (a)	Mass = volume × density = $2A\rho x$	1
10 (b)	Force = weight of extra liquid = $2Ax\rho g$ But for the whole tube of liquid $M = \rho AL$, so $F = M(2x/l)g$. The restoring force = $(2Mg/l)x$. $2Mg/l$ is a constant, so the restoring force is proportional to the displacement. Therefore the motion is simple harmonic.	2 2
10 (c)	Acceleration = force / mass = $(2Mg/l)x / M$ $= (2g/l)x$ The standard equation for acceleration in SHM is: $a = -\omega^2 x$, so $\omega^2 = 2g/l$, $\omega = \sqrt{(2g/l)}$ frequency = $2\pi/\omega = 2\pi/\sqrt{(2g/l)} = 2\pi\sqrt{(l/2g)} = \pi\sqrt{(2l/g)}$.	2
10 (d)	• Time a number of oscillations, 10 if possible. • Repeat this measurement a number of times and find the mean. • Repeat for different lengths of liquid. • Use a fiducial marker. • Plot a graph of frequency^2 versus length (or frequency *versus* $\sqrt{length}$). • Gradient will give $2\pi^2/g$.	4
		Total 11

Glossary

absolute temperature scale	temperature scale measured in kelvin (K), which has 0 K as absolute zero; 0 K = −273.15 °C; the other defining point on the scale is the triple point of water
absolute uncertainty	the size of the uncertainty in a measurement due to the resolution of the method of measurement; the uncertainty has the same units as the quantity being measured
absolute zero	the lowest possible temperature: the temperature at which there is zero kinetic energy of the particles (particles are stationary except for quantum-mechanical motion)
accurate	when a reading is close to the accepted value
activity	the activity of a radioactive source is the number of emissions per second; measured in becquerels (Bq); 1 Bq = 1 emission per second
alpha (α) particle	particle formed from two protons and two neutrons (a helium nucleus); emitted by the nuclei of some radioisotopes
alpha (α) radiation	short-range, highly ionising radiation consisting of helium nuclei
alternating current	electric current that is continuously changing in magnitude and direction, following a periodic pattern
amplitude	the largest displacement from equilibrium of an oscillating object, or the maximum height of a wave
angular speed (ω)	in rotation, the angle turned through per second; unit radians per second, rad s^{-1}
angular velocity	the rate of change of angular displacement with respect to time, given by $$\omega = \frac{\Delta\theta}{\Delta t}$$
atomic mass unit (u)	unit of mass defined as 1/12 of the mass of a carbon -12 atom
average molecular kinetic energy	the average molecular kinetic energy of an ideal gas is the average value of the mean translational kinetic energy of each molecule (vibrational and rotational kinetic energy are negligible in an ideal gas); it is given by the expression: $\frac{1}{2} m (c_{rms})^2$.
Avogadro constant (N_A)	the number of particles in a mole of a substance; $N_A = 6.02 \times 10^{23}$
Avogadro's law	law stating that equal volumes of gases, at the same temperature and pressure, contain the same number of molecules
background radiation count rate	the number of emissions per second that are detected due to radiation from the environment
becquerel	measure of radioactivity; 1 becquerel = 1 disintegration per second
beta (β) radiation	ionising radiation in the form of fast-moving electrons (or positrons)
binding energy (nuclear)	the energy required to separate a nucleus into its constituent protons and neutrons
binding energy per nucleon	the average energy required to remove each proton or neutron from a nucleus

Boltzmann constant (k)	a constant that links the absolute temperature of a gas to the average kinetic energy of its molecules: k is equal to the ratio of the molar gas constant R to the Avogadro constant N_A
Boyle's law	law stating that for a fixed mass of an ideal gas at constant temperature, the pressure of the gas is inversely proportional to its volume: pV = constant
Brownian motion	the observable random movements of particles such as smoke particles, caused by the high-speed thermal motion of liquid or gas molecules
capacitance	the charge stored per unit potential difference applied across a capacitor; unit farad, F
capacitor	a device that stores electric charge, and therefore electrical energy
centripetal acceleration	an acceleration towards the centre of a circular path
centripetal force	a force causing an object to move in a circular path; it acts towards the centre of the circle
change in state	the process in which energy is transferred to or from a substance, causing the separation between particles to change, e.g. from liquid to gas or from liquid to solid; there is no rise in temperature
Charles' law	law stating that for a fixed mass of an ideal gas at constant pressure, the volume of the gas is proportional to its absolute temperature: $\frac{V}{T}$ = constant
control rod	rod of neutron-absorbing material (such as boron or cadmium) which is used to slow the rate of fission in a nuclear reactor
coolant	a fluid used to cool a nuclear reactor core, most commonly water but sometimes carbon dioxide or molten sodium
corrected count rate	measurement of radioactivity that takes background radiation into account, so the corrected count rate measured is due only to the substance under investigation
Coulomb's law	the law describing the electrostatic force between two charges: the force is proportional to the product of the charges and inversely proportional to the square of the distance between them
critical mass	the minimum mass of fissionable material that can sustain a chain reaction
damped oscillation	oscillation for which the amplitude decreases with time as work is done against resistive forces
daughter nucleus	the nucleus that remains after radioactive decay has taken place
de Broglie wavelength	the wavelength, λ, of a particle when it behaves like a wave: $\lambda = h/p$ where h is the Planck constant and p is the particle's momentum
decay constant (λ)	the probability that a radioactive decay will take place in unit time; λ is equal to ln 2/(half-life), and has unit s^{-1}
dielectric	the insulating material between the plates of a capacitor
diffraction	the spreading or bending of waves as they pass through an aperture or round the edge of a barrier
direct current	electric current produced when charges drift in a steady direction

displacement (s)	a vector describing the difference in position of two points
driving frequency	the frequency of an external varying force which is causing oscillations
eddy currents	an alternating magnetic flux through an iron core induces emfs in the core, which drive eddy currents; these generate heat in the core resulting in energy wastage
elastic collision	a collision in which the total kinetic energy is conserved, i.e. the sum of the kinetic energies before and after the collision is equal
electric field strength	the force on a unit charge: $E = F/Q$
electric potential (V)	property of a point in an electric field: the work done in bringing a unit positive charge from infinity to that point; unit joules per coulomb, or volts
electron capture	during electron capture, an electron in an atom's inner shell is drawn into the nucleus where it combines with a proton, forming a neutron and a neutrino; the neutrino is ejected from the atom's nucleus
electron volt (eV)	unit of energy, equal to the energy transferred when an electron moves through a potential difference of 1 volt; 1 eV $= 1.6 \times 10^{-19}$ J
enriched nuclear fuel	nuclear fuel that has had the proportion of its fissionable isotope artificially increased; for example, uranium that has had the proportion of U-235 increased
equation of state	an equation describing the state of matter under specified conditions
equation of state for an ideal gas	equation linking the bulk properties (pressure, p, volume, V, and absolute temperature, T) of n moles of an ideal gas: $pV = nRT$
equipotential surface	a surface where the potential is the same everywhere: for example, in an electric field no work is done moving a charge along an equipotential surface
escape velocity	the minimum initial velocity needed at the surface of a planet or star if an object is to escape its gravitational pull; on Earth an object thrown upwards at 11 km s^{-1} would totally escape the Earth's gravity and never fall down again
exponential decay	the reduction in magnitude of a quantity by a certain factor, e.g. half, in a constant time period; the activity of a radioactive source follows an exponential decay
farad (F)	the unit of capacitance; a 1 F capacitor would store 1 C of charge for every 1 V of potential difference applied across it
Faraday's law (of electromagnetic induction)	when there is relative movement between a conductor and a magnetic field, the e.m.f. induced in a conductor is equal to the rate of change of magnetic flux linkage
femtometre (fm)	unit of length of nuclear dimensions; 1 fm $= 10^{-15}$ m
fiducial marker	an object used as a marker to make it easier to measure time intervals for a pendulum bob passing a reference point in its swing
field lines	visual representation of the field of force around the object producing the field; the direction of the field lines indicates the direction of the force produced on a positive charge or mass; the separation of the field lines represents the strength of the field
first law of thermodynamics	a law stating the conservation of energy – the change in internal energy of a system (a solid, liquid or gas) is the sum of the energy transferred to it by heating or when work is done on it by an external force.

fission	the decay of a some large nuclei by splitting into two smaller nuclei, accompanied by a release of energy
flux	total magnetic field passing through a particular area; magnetic flux = flux density × area
flux linkage	within a current-carrying coil, the sum of the magnetic flux due to each loop of the coil
forced vibration	an oscillation caused by an external varying force
free oscillation	one in which a body or system oscillates at its natural frequency
free vibration	simple harmonic motion in the absence of any external, varying force
frequency	the number of oscillations or waves in one second; unit hertz, Hz.
fusion (nuclear)	the formation of a larger nucleus by combining two smaller ones, accompanied by a release of energy
gamma (γ) radiation	ionising radiation emitted during the decay of some radioisotopes; high-energy (very short-wavelength) electromagnetic radiation
gas constant	*see* molar gas constant
gas laws	the fundamental relationships between pressure p, volume V and temperature T of a gas under specified conditions
geostationary satellite	a satellite in an equatorial orbit over a fixed point on the Earth's surface; its orbital period is 24 hours
gravitational field strength (g)	the force on a unit mass at a point in a gravitational field; unit N kg^{-1}
gravitational potential (V)	property of a point in a gravitational field: the work done in bringing a unit mass from infinity to that point; unit joules per kilogram
gravitational potential difference	gravitational potential difference, ΔV, is the change in energy, ΔW, of a body of mass m caused by a change in position within a gravitational field: $\Delta W = m\,\Delta V$
ground state	an atom is said to be in its ground state when its electrons all occupy the lowest possible allowed energy levels
half-life	the time taken for half the nuclei in a sample of a radioisotope to decay; *or* the time taken for the activity of a radioactive source to drop by half
heat capacity	the energy required to raise the temperature of an object by 1 K; unit J K^{-1}
ideal gas equation	equation of state linking the bulk properties (pressure, p, volume, V, and absolute temperature, T) of n moles of an ideal gas: $pV = nRT$
ideal gas	a gas that obeys Boyle's law under all conditions: a gas whose molecules are infinitely small and exert no force on each other, except during collisions
induced e.m.f.	a potential difference across a conductor in a complete circuit, caused by a change in the magnetic flux around the conductor
induced fission	fission of a nucleus, usually uranium-235, caused by the absorption of an extra neutron
intensity	the power of a wave transmitted through unit area perpendicular to the direction of travel of the wave; unit watts per square metre, W m^{-2}

internal energy	the energy of a substance due to the sum of the kinetic energy and potential energy of all its constituent particles
inverse square law	law describing how radiation, gamma radiation for example, spreads out in three dimensions from a point source; the intensity of radiation is inversely proportional to the square of the distance from the source: $I_2/I_1 = (r_1/r_2)^2$
ionisation	the process of removing electrons from atoms
Kelvin scale	*see* Absolute temperature scale
kinetic theory	theoretical model of a gas that explains its bulk properties, such as temperature and pressure, in terms of the motion of its molecules
latent heat	the energy needed for a substance to change state, with no change in temperature
Lenz's law	the direction of an induced e.m.f. is such that it opposes the change of magnetic flux that caused it
log–linear graph	a graph with a logarithmic scale on one axis and a linear scale on the other axis
magnetic flux (Φ)	the magnetic flux through an area A is equal to (flux density perpendicular to A) $\times A$; unit weber, Wb
magnetic flux density (B)	a measure of the strength of a magnetic field; the flux density is equal to the magnitude of the force on a unit length of current-carrying wire per unit current; unit tesla, T
magnetic flux linkage	the total magnetic flux linkage through a coil of area A and number of turns N is $N\Phi$ or NAB
mass difference	the difference between the mass of a nucleus and the total mass of its constituent nucleons; sometimes called the mass defect
metastable state	an excited state of a nucleus with a relatively long half-life
moderator	material used in a nuclear reactor to slow down neutrons without absorbing them, e.g. graphite
molar gas constant (R)	the constant of proportionality in the ideal gas equation; $R = pV/nT = 8.31\ \text{J K}^{-1}\ \text{mol}^{-1}$
molar mass	the mass of one mole of an element or compound
mole	the amount of substance a system contains in terms of the number of atoms as there are in 0.012 kg of carbon-12
molecular mass	the mass of one molecule of a substance; $\text{molecular mass} = \dfrac{\text{molar mass}}{\text{Avogadro constant}}$
momentum (p)	property of a moving object equal to its mass (m) multiplied by its velocity (v), $p = mv$; a vector quantity, unit kg m s^{-1}
motor effect	a wire carrying a current in a magnetic field experiences a force; this is known as the motor effect
natural frequency	the frequency at which a vibrating object undergoes free vibrations; for example, the frequency at which a tuning fork will oscillate when struck

Newton's law of gravitation	the gravitational force between two point masses is proportional to the product of the masses, and inversely proportional to the square of the distance between them
oscillation	a repeating to-and-fro motion, such as that observed in a swinging pendulum or vibrating guitar string
oscilloscope	device for measuring alternating signals e.g. alternating current
parent nucleus	nucleus of a radioisotope that emits radiation and decays to the daughter nucleus
parabola	a curve with the general equation, $y = ax^2 + bx + c$; the trajectory of a moving charged particle entering a uniform electric field initially at right angles is a parabola, as is the path of a projectile under the influence of gravity
peak-to-peak value	the maximum displacement across both directions of an alternating current
peak value	the maximum displacement from the zero line in either direction of an alternating current
percentage uncertainty	(uncertainty / measured value) $\times$ 100%
permittivity	the ratio of the electric displacement in a material to the intensity of the electric field producing it; in free space it is 8.85×10^{-12} F m^{-1}
permittivity (of free space)	the permittivity of a medium is a measure of the material's ability to resist the formation of an electric field within it; using a dielectric with a permittivity greater than that of free space (a vacuum) increases the capacitance of a capacitor
period (*T*)	the time taken to one complete oscillation, or one full rotation
periodic motion	a repeating pattern of motion, such as rotation or oscillation
polar molecule	a molecule such as water which has a slightly positive part and a slightly negative part
potential difference	the (electric) potential difference between two points is the energy transferred per unit charge moving between the points
power	the rate of doing work
precise	precise measurements are ones in which there is very little spread about the mean value
pressure or **pressure–temperature law**	law stating that for a fixed mass of an ideal gas at constant volume, the pressure of the gas is proportional to its absolute temperature: $\varepsilon = 1 - \frac{T_C}{T_H}$
probability	measure of the likelihood of a radioactive nucleus decaying in a certain time; for a large number of nuclei N over a certain time Δt, the number of nuclei that will decay ΔN is given by probability $= \frac{(-\Delta N/N)}{\Delta t}$
radian	a unit used to measure angle; 1 radian is the angle subtended at the centre of a circle by an arc whose length is equal to the radius; 2π radians = 360° 1 rad = 57.3°
radioisotope	a radioactive isotope

relative permittivity	the factor by which the presence of a dielectric raises the capacitance of a parallel plate capacitor, relative to a vacuum; also called the dielectric constant
resonance	large amplitude vibrations caused when the driving frequency matches the natural frequency of a system
resolution	the resolution of a measuring device is the smallest increment in the measured quantity that can be shown on the device
resonant frequency	the natural frequency at which an object or system oscillates
restoring force	a force that acts so as to return an object to its equilibrium position
root-mean-square (r.m.s.) speed (c_{rms})	the square root of the mean squared speed (of a collection of particles)
rotational frequency	the number of rotations per second; unit hertz, Hz (or sometimes revolutions per minute, rpm, but this is not an SI unit)
Rutherford scattering	the elastic scattering of charged particles such as alpha particles by similarly charged particles such as nuclei
scalar	a physical quantity that is fully specified by its magnitude (size); it has no direction associated with it
search coil	a flat coil of insulated wire with a large number of turns; when connected to an oscilloscope, the search coil can be used to measure the strength of a varying magnetic field, since the emf induced in the coil when placed in a varying magnetic field is proportional to the flux density of the field
simple harmonic motion (SHM)	periodic motion such that the restoring force is proportional to the displacement from equilibrium and acts in the opposite direction
specific heat capacity	the energy required to raise the temperature of a 1 kg mass of a substance by 1 K; unit $J\ kg^{-1}\ K^{-1}$
specific latent heat of fusion	the energy needed for 1 kg of a solid to change to a liquid, with no increase in temperature; unit $J\ kg^{-1}$
specific latent heat of vaporisation	the energy needed for 1 kg of a liquid to change to a gas, with no increase in temperature; unit $J\ kg^{-1}$
stationary wave	a wave that does not transfer energy in the direction of wave travel
synchronous orbit	an orbit in which an orbiting body (such as a satellite) has an orbital period equal to the rotational period of the body (such as a planet) it orbits, in the same direction of rotation as that body
thermal neutron	a slow moving neutron which can be capture by a fissile nucleus, causing nuclear fission
thermometric property	a property of a substance that changes with temperature, such as the volume of a gas, the length of a mercury column or the electrical resistance of a wire; can be used to provide a temperature scale
tesla (T)	unit of magnetic flux density; 1 T is the magnetic flux density when 1 m of wire carrying 1 A of current at right angles to a magnetic field experiences a force of 1 N

time constant	time taken for the charge on (or voltage across or current flowing off) a discharging capacitor to drop to 1/e (≈0.37) of its original value; it is equal to CR, where C is the capacitance and R is the total resistance of the circuit; unit seconds
time period	the time taken for one complete cycle of an oscillation
tracer	a radioisotope, usually a gamma emitter, used to track the passage of a substance; for example, radioactive iodine is used in the body to monitor thyroid function
transformer	a device that uses electromagnetic induction to change the voltage (and current) of an a.c. signal
transformer equation	$N_s/N_p = V_s/V_p$
uncertainty	a measure of the precision of experimental results; may be expressed as a percentage of the value
vector	a physical quantity that is specified by its magnitude (size) and its direction
wave–particle duality	the concept that all energy (and matter) exhibits both wave-like and particle-like properties
work done	the product of the force and the distance moved; for a gas at constant pressure it is given by $W = p\,\Delta V$
weber (Wb)	the unit of magnetic flux; 1 Wb = 1 T m^2

Index